Responses to the original 'Theory of Involution'

'At long last I have read your typescript from A to Z. Needless to say I agree with much, even most of what you have to say. This implies an answer to your question whether it is right to rewrite and expand your theory. It would be definitely worth it. I feel less optimistic about the possibilities of publication.'
Arthur Koestler, 1978.

'I thank you very much for sending me your Theory of Involution. I have just begun to read it and it interests me enormously.... I certainly do share your views ... and believe, like you, that so called evolutionary progress is explicable in scientific terms. I shall write again...'
Prof Konrad Lorenz, 1970. Nobel Prize 71 (Max Planck Institute, Bavaria)

'Thank you so much for your letter and your manuscript on the "Theory of Involution"... I have had a good look at it and it seems you have a great deal to say that is of vital importance right now....'
Irwin Schumacher, 1974, (Author of *Small is Beautiful: a study of economics as if people mattered.*)

Praise for 'Involution- An Odyssey Reconciling Science to God'

'A brilliant and profoundly erudite epic charting the evolution of Western thinking processes, probing the frontiers of rationality and naturalism and opening up a deeper understanding of the nature of reality based on the reality of mystical experience. The author's grasp of the principal elements of Western culture is masterly and her poetic narrative woven together with extraordinary subtlety. The detailed footnotes demonstrate a rare depth of perceptive scholarship. This is nothing short of a heroic intellectual tour de force and deserves the widest readership.' (David Lorimer, Director Scientific and Medical Network)

'Philippa Rees wrote a book that is a rarity: it is on a controversial, actually hair- and eye-brow-raising subject, and it is totally sincere. And totally insightful. If you the reader are as brave as this author, you are in for a fantastic ride. Getting close to science as well as to God at the same time. That's no mean feat. Enjoy the ride – and the light!'
Ervin Laszlo

'...*Your journey through poetry is more than just an alternative treatment of the material you originally theoretically described; it is the very act of genius, which is able to treat the ambiguous nature of the world differently. The poetry is an alternative for how the world makes meaning from the ambiguous. It is a completely alternative direction for an exploration of the world in itself.*

The scale of the feat you have thereby achieved by writing in poetry is immense. This goes far beyond the mechanistic notions of wholeness arrived at by some modern scientific authors. Your work reintroduces the aesthetic, beautiful, meaningful process that is poetry into science. The genius of involution is not just a mechanism of science relating to the whole but a completely different realisation of the beautiful, within living process.'
(Philip Franses. Editor. The Holistic Science Journal. Lecturer Schumacher College)

"*Involution* is, at *least in terms of 'subject', a daring, Dantean feat.*

Rees's profound notion that the evolution of humankind is made possible by the dormant dominions of evolutionary memory in our unconscious - the eponymous 'involution' - is, I would suspect, a theory Charles Darwin would have gratefully embraced as a curative to his own bleaker 'discoveries', which he initially emotionally and religiously resisted.

That Rees has chosen to communicate her dialectic in the medium of sprung-rhymed blank verse is ingenious in itself, as well as being in the narrative spirit of the poetry of the ancients.

Whatever one's poetic, religious or scientific response to Involution may be, what will be difficult for even the most scouring of critics to deny is it's scholastic vitality, compositional discipline and macrocosmic scope. Involution is a work of indisputably tall ambition, and an accomplishment which may well prove much more than the sum of its invariably exceptional parts".
(Alan Morrison. Poet. Editor: The Recusant On-Line Magazine)

INVOLUTION
An Odyssey Reconciling Science to God

P. A. Rees

Published by
CollaborArt Books

Prose Editor: Karin Cox
Cover design: Ana Grigoriu
Back Cover Image: Z Thomas
Interior Design: Shore Books and Design
Internal Diagrams: Philippa Rees

First Print Edition 2013

**Published by:
CollaborArt Books.**
(collaborartbooks.com)

Copyright© Philippa Rees 2013

ISBN 978-0-9575002-2-8

All rights reserved. No part of this publication may be reproduced, stored in a retrieval system or transmitted, in any form or by any means without the prior written permission of the author, nor be otherwise circulated in any form of binding or cover other than which it is published and without a similar condition being imposed on the purchaser.

The footnotes to this work provide full credit to all quotations and interpretations mentioned in the text. Every effort has been made to contact copyright holders and obtain permissions; nevertheless, the author welcomes correspondence from copyright holders wishing to suggest amendments in subsequent printings and editions.

Printed by Lightning Source

We are not human beings having a spiritual experience; we are spiritual beings having a human experience.
Père Teilhard de Chardin

For many silent companions: those who brought gifts but mistook the mood of the party, arrived too early and left unsung.

Acknowledgements

In considering those who have lived with the various incarnations of this theory—its hopes and despairs, and its refusal ever to take second place—first tribute must be paid to my husband John, who has eaten more indifferent food in distracted silence than should be asked of any man. He has never wavered in his support for a writer unlikely to be published, with an idea no one wanted. Heroism and stoicism in him have been indistinguishable. My daughters, Juliet and Kate, have mostly uncomplainingly accepted sharing a mother with an obsession. Juliet has read, offered opinions, and re-read. Kate has waited patiently until it might all be over. Both have provided balanced weight to the pivot of sanity, and given other things to compel attention. I can never repay them for their generosity.

To them must be added tolerant friends who seemed, without good reason, to believe in this work. First tribute and gratitude is owed to Sophie Wilkins, who read for a living and read my work out of kindness, despite my scripts 'balancing badly on the belly' in bed. She believed in Involution when no one else did, and sustained my belief through very lean times. To Joanna Cherry, who kept intermittent vigil but admitted that, after forty years, it was a relief to be able to broach it 'in a mixed assembly'; Kristin Yates, who accepted section after section attached to emails and cheerfully used a virtual blue pencil before returning them; Jen Kershaw who took it running from a cold start and never missed a beat.

To them, I would add Dick and Margaret Milford, much missed, who in different ways (and for different reasons) supported this idea by offering a bean-covered caravan and food for long periods of destitution, and long conversations on Teilhard de Chardin over the washing up. Their kindness to an unknown homeless immigrant was both constant and extraordinary; the light in my life for many years.

The original support of scientific philosophers Arthur Koestler, Konrad Lorenz, and Irwin Schumacher has had to do duty for all the intervening years. To them I owe sustained conviction, which had to be eked out after their deaths. They exemplified the generosity of the genuine academic in those days, who never questioned one's qualifications to hold an idea—just as well. Finally, this work is written in gratitude for the experiences that gave rise to the idea itself. The vision that led to it was never easy to integrate, but it has sustained a belief that nothing is without meaning and thereby enriched all of life. I do not know why I deserved it.

Preface

This work is not what it may at first appear – a quirky or contrived way to tell the familiar scientific story. Instead, it offers a journey, not of instruction or persuasion, but of discovery. The language of poetic suggestion, less domineering than prose, constructs a ladder from familiar material, but only to afford a panoramic view of what lies beyond science, and all languages: the laying down and recovery of memory, and memory's crucial role in the processes of evolution. It is the enfolding and recovery of memory that this book descends inwards (and thereby upwards) to expose.

To list the constituents of the ladder (the scientific facts and arguments – merely the necessary vocabulary) as 'contents' would be to misrepresent the book, and obscure its real purpose, which is simply to provide a wider and alternate view. Neither poetry nor science on their own could achieve this, but somewhere between them a new vision might take shape, rather as music carves melody from silence, and having changed everything, returns to it. Involution, the in'folding' of memory, and its recovery by man, does not seek to challenge Darwin, whose world is the evolution of outer forms, but offers a counterbalance, the in'forming' of evolving awareness that shapes his creatures and their relationships.

The central premise of this book is that evolution, prior to man, has been achieved through the infolding of experience – involution – and that its legacy, memory, resides in the very structure of matter. After the emergence of man, this continues at a higher level in the building of the conscious mind. Involution, in man, continues through the recovery of memory by incremental inspiration, and it builds the scientific model of that memory; the collective intellect. Since instances of inspiration are individual, never repeated, and cannot be validated, it does not therefore qualify as a scientific theory. Not yet.

As inspiration is momentary, spontaneous and wordless, excluded from legitimate scientific enquiry, the only evidence for this hypothesis is the entire sweep of human history. Only a high perspective reveals the parallel between the successions of increasingly complex creatures and their sequencing through time, (evolution) and the reverse chronology of science's recovery (higher level involution). Scientific disciplines emerge in answer to the demands of that recovery. From the comprehensive cosmologies of the ancients, through the fracturing of specialist division, back towards modern universal theories of the cosmos, the unbroken continuity accelerates in the last hundred years, just as evolution did in its convergent ascent to Man. The furious pace of technology is the debris of that hectic recent recovery, and the approach to origins.

The mirror that involution offers to evolution provides compelling evidence of a collective process which implies an underpinning law—not a law imposed, but a coherent law that evolves. They are two sides of a single coin, matter-mind. In time, science may accept the inevitability of extending its legitimacy into the realms of consciousness. Not brain waves enclosed in skulls but the structure of the shared field of consciousness in which all forms partake and which brains interpret. All this premise can offer by way of evidence is the pattern of recovery and the circumstances conducive to scientific inspiration and inspiration's echoes of mystical experiences. Both point towards the same creative truth.

This building of the ladder, or what might be more appropriately and helpfully visualised as a single collective DNA helix, begins with pre-human molecular and cellular memory and rises to the expanding complex consciousness that has afforded Man a view of everything except perhaps the compelling rope of his climb. In the company of Reason and Soul, who between them support the opposing mirror coils, this journey is about the climb, through those maverick geniuses who, through visionary syntheses, set the unifying connecting steps for all to follow, and one after the other, shimmied up the centre. The unerring thread of memory drew and linked them, from Omega to Alpha, from the Hellenic Aegean to the Mverse and the Large Hadron Collider. It has been a single journey, collectively achieved, collectively paced, and lit by the torch bearers.

The Table of Contents below roughly lists the epochs through which this journey travels, and some themes that distinguished each, but the details wait to be encountered. Every traveller will perceive and orientate differently. Where you start from very much determines the route likely to be taken. It is best not to over-define what may never be agreed upon. Intuition and liberty are central to real understanding but I hope the travel, in itself, will divert and be worthwhile.

Table of Contents

Introduction   1

    Where we are now: A broad over-view of evolutionary ideas, and what difference involution makes. Why poetry best serves its evidence.

Canto the First:
The Map and Older Tracks   19

    This reviews the past traditions of religious revelation and poetic inspiration, and the gulf that opened between rational and intuitive understanding. Early science and poetic odysseys were both shaped by envisaged perfection: holistic integration, or the absorption and paradise afforded by women and love; in contrast to the limits of reductionist analysis. Art comes closer to truth.

Canto the Second:
Memory, Dreams and Separation   37

    Some overall patterns constant throughout: Positive and negative oscillation: Complexity arising through synthesis, and the developing 'inner selection' prompting spontaneous maverick action and acceleration.

Canto the Third:
The Earliest Days, Society and Language.   63

> Early man, society, tools and language: the microcosm of the macrocosm, and the beginning of the divided mind and the separating of intellect.

Canto the Fourth:
Greece and Some Special People.   79

> Recorded time: early ideas contributed by fertile genius, the unitary beginnings, the whole cosmos, and the start of divergence. The fracture between Plato and Aristotle cleaves that future division, between religion and science respectively, for all time.

Interlude: Archimedes and Alexandria.   99

> A moment of respite: in which the companions are faced with Archimedes' rejection and his analysis of the limits of intellectual endeavour as a path to truth.

Canto the Fifth:
The Dark Ages, Monasteries and Muslims.   107

> Rome accumulates. The Church dominates. Celtic Christians: Monastic preservation of Greek sources: The Islamic world of chemistry and mathematics. Controlled artistic expression and prevailing ideas.

Canto the Sixth:
The Renaissance and Liberty.   131

> Explosive concurrent developments: Geology in Britain: Anatomy and perspective in Italian painting. The exhibits

of diversity are still subject to concepts of perfection: Pascal, Gilbert, Galileo and Kepler, followed by the rationalism of Descartes defining the limits of mind and matter for all future science.

Canto the Seventh:
The Enlightenment and Rationality.   159

The fully liberated intellect concerned only with the 'outer' and material: Newton and mathematics as the scientific language. New disciplines created by single thinkers, Boyle, Hooke, and Dalton delimit increasing specialisation as the mind reflects the diversity of recovered evolutionary memory. Finer subdivisions start to uncover the underlying unity. Classical painting and music provide a mirror to the formal analysis of science.

Canto the Eighth:
Modernism and Dissolution.   185

Involution returns to the origins, the Serengeti, where human thought began. The Kilimanjaro of excavated memory starts to melt; as evidence the dissolution of modern art and music. The pattern of either/or oscillation is danced through alternate chronological concepts; electricity and magnetism become united through ever widening law of forces and fields, matter and energy, ending with the last of the irreconcilables, relativity and quantum theory. The complete division between mind and matter are now the consequence of the complete separation of intellect from consciousness, science from religion, man from the created world, and from his deeper nature.

Canto the Ninth:
Love and Reunion.   209

Soul continues alone. Experience is the solitary constant we have traced through maverick leadership, the connecting steps of the ladder. This soliloquy of the serpent reminds us of the journey, now traversed, the slender walkway built by Reason's historical narrative. She then reveals her structural relationship with mathematics, geometry, with time, and the intuitions of genius to whom the storage of evolution's memory has been, in fragments, revealed. The serpent is the gatekeeper, mediating between the world of mind and matter. Science has uncovered hidden memory, and built a model, a left brain intellect, but the cost has been alienation, man from his true nature, in unity with all.

Women are, at the end as at the beginning, the means of life and the idols of romantic aspiration, the practitioners of death in every birth, through which Man makes his ascent and his return. Through memory he has, collectively, come full circle. All that remains is the individual experience of love; dissolving all separation and silencing all words.

Appendix: 237
Saints and Scientists and the common threads that link them.

Afterword: 251
The experiences that led the author to this vision.

Footnotes to all Cantos 261

Bibliography 415

Introduction

Over the past decade, many weighty, erudite books have been written seeking to make good the deficiencies of Darwinism; slimmer ones have offered strategies to compensate for the theory of evolution's barren landscape. Science is running out of matter, mankind out of hope, and mind is hardly to be seen. These books propose a variety of solutions. Ervin Laszlo's Akashic field theory and Rupert Sheldrake's concept of morphic resonance supply ideas to introduce mind or memory's influence in some form. String and M-theory supply alternatives to make good the deficiency of matter and energy. All provide intellectual hypotheses.

This work is different. Essentially, it is two books: the first a poetic odyssey through the chronology of inspiration; the second, footnotes that explain the speed, content and connections of science. The marriage between them offers nourishment of different kinds. Some readers will prefer one over the other; some will seek a balance between them. The overarching process could be expressed equally validly using the history of philosophy, art, or any sphere of language, for all disciplines of thought reveal the same journey. The journey is the odyssey through memory—not only the memory of the single life, but through the life of all creation and the creation of all life—in a word, evolution. Any thread pulled from the tapestry would reveal the same patterns. Choosing the spectrum of science was a conscious decision, because science is a process and not merely a body of thought. It changes constantly, building chronologically, and therefore it is easier to trace science's oscillations through time. Yet the choice will lay me open to charges of simplistic naïveté, inaccuracies (no doubt), and presumption. No matter. It is entirely the process I wish to convey; the facts, or even the ideas, are subordinate. Where and to whom they came, and when, and how, is all that matters.

This approach seeks to counterbalance science's belief that intellectual solutions are sufficient. Or, at this point in time, that they are of prime importance. Undoubtedly, science needs invigoration and new directions, but it is not only minds that need changing, but hearts. That is not achieved by ingenious argument or plausible hypotheses. The sole and unequal reliance on the intellect is arguably the most destructive contribution made by modern science, however successful its technology. It has steadfastly ignored what might have integrated science within creation itself and might have led to wisdom, rather than just to knowledge. Science is responsible, not for what it has attended to but for what it has wilfully ignored—its own origins—and for its habit of refusing anything new that might challenge its complacency. Creation has paid a high price for science's limited certainties.

What this work seeks is not to belittle the considerable achievements of science, but to realign them. It seeks to recover the critical, unique contributions of individual subjective experience—numinous experience, revelation, inspiration, and intuition—and the explosive energy they contributed throughout evolution. More tellingly, it seeks to show how they contributed throughout human (including scientific) thought. The encounter between the individual and the spiritual, in its broadest sense, has shaped theory and enriched life. These encounters are not susceptible to scientific verification, are seldom repeated, are unique to the individual (for whom they provide critical understanding), and through them, achieve consequent renewal for all. Great minds may not only think alike, they may do the thinking for the rest of us.

This book traces the way in which genius, always in the vanguard, has pursued the recovery of memory—the memory of evolution. That recovery is involution, but it does not arise with man; it is the continuation of the means by which creation happened at each moment, through acts of self-transcendence, experimentation, or spontaneity. In man, the higher act is in the mind, both individual and collective, continuing patterns established in the instincts of animals and the insights of our close mammalian relations.

For this reason, I have purposely chosen poetry to avoid an intellectual argument or hypothesis, and closed the work with an invitation for readers to be open to such direct experience. Closing the wrong door is often necessary, preliminary to recognizing the door that remains ajar. This is a science of, and for, the individual, and it is with the individual that I chose to close it. Its wider relevance as a theory, if recognized, I happily entrust to others.

Through re-examining how gifted individuals made their enriching contributions to the nobility of mankind, I hope to lead the reader towards a new way of knowing where and how to look. Thereby, the silent growth and influence of consciousness might return spiritual man to his origins, and with him, propel integrated creation to a new, yet unfamiliar turn in the spiral. Hence, human strife and environmental chaos; all waits upon all. The dissolution of matter's distinctions is mirrored in the dissolution of separateness—the growing global embrace, against which exclusive '-isms' (nationalism, fundamentalism, separatism) are all fighting like tigers. Correction: fighting as only man can.

Although this seems to be a journey through the history of science, that journey serves a more important purpose: to present an entirely new evolutionary theory—one that makes mankind inheritor and custodian, rather than apogee or terminus. The theory of involution proposes that consciousness has, through the forms of material and biological change, been the driver of evolutionary advances. Through such advances, and their interactions with each other, consciousness itself has co-evolved. Thus, creation in its multifarious forms partakes of and expresses consciousness. Creation is a coherent whole—each part, each form, is both individual and unique, and a participant in the evolving and involving development of consciousness. It has always been so, but now that recognition has become imperative.

This hypothesis, therefore, is less theory than *theoria* (Greek, 'contemplation' or 'sight'). It was not derived from deductive reasoning or an amalgamation of ideas from other books. It was lit by the tungsten flare of uninvited experience, which could not be encompassed within the limits of scientific understanding nor be considered legitimate for scientific examination. Following those experiences, I realized that the dimensions of creation were so much grander, so much more focussed than Darwin knew or could have known. The evidence for this alternative required all the disciplines of science. No collection of books, no single discipline, and no area of expertise could have furnished the theory of involution at that time. Certainly, once seen, it proved to be widely suggested by almost every book I encountered—all the products of human thought and accounts of spiritual experience lay along the continuum of intermittent and collective recovery. None have provided this theory, but equally, none have contradicted it; almost all have added coherence and detail. I plead impartiality and innocence as the precondition of understanding.

It is a convention of science to claim deductive, *a posteriori* process. This, as all imaginative scientists know, is nonsense. The hypothesis governs what is sought and observed—what it is designed to find. The Large

Hadron Collider was not built in case it might prove useful; it was built to find, and what it will find will be already 'inbuilt' in its search. There will be enough alternatives in how particles collide for one or another manifestation to sign a QED, even if it is not the one hoped for. The evidence for involution, on the other hand, is genuinely deductive—an unselective deduction encompassing everything. Although its inspiration was a 'eureka moment', what the experience illuminated was the nature of all of those other seminal eurekas, floating on the surface of history in a chronology that curiously mirrored, exactly, the chronology of evolution. It was as though man's collective mind lay at the service of memory, and man's memory was integrated with everything that had preceded him. Genius merely uncovered the next cave in the underground passages that connect us all, and dropped the paper to chase. The languages varied, and approximated, but none defined, nor could any one expropriate an explanation, each merely revealed a universal pattern.

Science, in both its contents and chronology, retraced the path of evolution in reverse: from man, all of man in Egypt and Hellenic Greece, to all of creation in relativity, Omega to Alpha. Did that make the inspired scientific genius no different from his religious counterpart? Was the nature of inspiration separated only by language and consequence? Was something as obvious as that separation responsible for wars and the destruction of the globe? It seemed too simple, but its coherence would not go away.

The compulsion to reconcile the great intellectual divisions that imprison thinking and prevent communication, that render science blind to its origins and intolerant of appeals to values or purpose (and the equal deafness of the religions to its validity, limited though it may be), has taken a lifetime.

The first draft of my 'Theory of Involution' was written in 1970, in a white heat as a condensed scientific monograph for journal publication. It failed. A request by Cambridge University Press to expand it for book publication was withdrawn after the first two chapters. It was 'too broad for general science and too scientific for the general'. Fifty copies posted to top scientists in every field were ignored. Only three scientific philosophers supported it—in that they supported one of the theory's central tenets: the collective nature of consciousness and to what it consents. Due to the need for consent all agreed it would be unlikely to find a publisher. One scientist alone stopped to pen a diatribe (and then proceeded to collect the evidence and was heralded, when he died, as a brave pioneer!). At that time, involution sought to stopper the bottle before the genie escaped. This odyssey seeks, with the hope of companions, to recapture it.

As science, it was rough-hewn—as all new theory is. It limited itself to those areas for which there was experimental evidence, mostly in the animal kingdom and in comparative animal behaviour. It remains cobbled together for want of other minds to dovetail and refine it, but increasingly, the experimental and conjectural gaps it leapt over forty years ago have been filled by neuroscience examining brains, medicine witnessing miracles of healing, by the study of information in stem cells, and by anthropology in talking to shamans. In addition, psychotherapists are now listening to clients about near-death and out-of-body experiences, and about past-life memory. The diversity of humanity's experience of consciousness is being garnered, as Darwin on the HMS *Beagle* once garnered specimens. The continuity of consciousness that underlies and connects all life will soon flow unimpeded by scepticism and doubt. Belief determines what is sought for; alter belief, even a little, and change 'never' to 'possibly' and new water trickles, for believing is seeing.

Since writing about involution, and in the period of withdrawal that followed its utter rejection, I have recently encountered the work of others that echo and detail many of the theory's suppositions. While this work may still seem 'suspect' to orthodox science, I now have good company in Wilbur, Laszlo, Grof and many others. The groundswell of such thinkers was predicted in the theory of involution over forty years ago, and I have been expecting them long hence. Is my contribution now superfluous? I did wonder, but there remain aspects of my original theory not repeated or considered by these more analytical theories, so I decided to add a shoulder to the wheel for a different readership—one unfamiliar, perhaps, with even the simplest of science and possibly perplexed by its soulless consequences and its legacy of panic for the wellbeing of our beautiful globe. In tracing the journey, I sought the renewal of hope—not in emergent hypothetical theory, but in the deeper, wider continuation of invisible constants that are now being noticed by some scientists and philosophers. Contemplatives have acknowledged them always.

Therefore, to attempt this alternative in poetic form was not a capricious decision. Shifting the weight of collective reality while retaining its factual validity is like single-handedly shifting a Chesterfield sofa. No sooner do you lever up one end then the weight of it slips away, and it remains as it was, in the same dogged position. To examine the familiar from a new perspective, when that familiar is the very foundation on which science rests and casts its shadow, requires a new light, a new way of apprehending. Changes in thought have always required new ways of looking, and having been seen, new ways of expressing. The prose of science is weighted with the legacy of the past. Words have narrow meanings. Concepts have restricted application. The liberty of metaphor (*meta* = across; *pherein* = carry) permits this theory to carry the argument across the familiar trodden mud toward a fresher field to plough.

It may be obvious, but a theory that explains consciousness as the hand on the reins of creation implies a continuity between the created forms that preceded man's emergence, as well as the evolution of mankind from his hominid forebears to the most advanced thinkers of the past three thousand years. The involution of collective human thought continues the involution of forms at a higher level, where thought is the higher 'act' by which the thinker encounters consciousness and by which consciousness finds its thinker.

The history of human thought is punctuated by what is commonly called genius. Geniuses play a large role in this theory as the emissaries of consciousness. Consciousness speaks to the collective mind through them, erupting in their ready minds when the collective is ripe for advancement. In our celebrity-obsessed culture, genius has overtones of individual isolation, remote inaccessibility, freakishness and fortuitousness. It is part revered, part resented. The original Greek etymology of the word *genius* is that of the 'attendant or tutelary spirit allotted to every person at birth', and this hypothesis of collective consciousness finding the natural talent or constitution to express its involution, recalls that original meaning. 'A man of genius … is a spring in which there is always more behind than flows from it', wrote James Froude. Froude's quotation is apposite to both the role ascribed to genius and to the individual genius's dispensability. If, having been found, the genius fails to find a way to articulate his inspiration, another will be found instead. Consciousness has its own impetus and does not stop for fallen warriors.

The synchronous eruption of groundbreaking ideas in different people at the same time, across continents and without contact, is frequently recorded, as is the revisiting of such ideas by later men and women in more welcoming times, and the desolation (and often insanity) of those who died unsung. The susceptibility of certain minds, and what made them so, is as interesting as the ideas they promulgated. Idealism hatches in barren places and in affluent ones. One thing characterises genius: the ability to unerringly recognise and locate the essence of originality, or to stumble upon its pre-existence. This work retraces the continuity running through them, the thread of consciousness finding, like water, the downhill open channels.

There is a paradox in offering involution to a world of such emphatic individualism: individualism but little individuality. Evolution, through the concept of the selfish gene, renders man's view of himself as a disposable vehicle with death as his end, life without purpose. Modern man is no longer aware of his individuality, but conscious only of his insignificance. If insignificant, why should he be moral? Or self-denying? By contrast, through involution the individual is perfect in himself and responsible for himself, and his genius lies in that uniqueness. The geniuses history celebrated were those who knew that truth intuitively (and compulsively) and lived and acted upon it, finding creative languages to convey the universal; hence mankind's recognition of their truths. Involution explains that recognition.

Evolution: Current Thinking

For the purpose of this book, I summarize the mainstream neo-Darwinist viewpoint despite intermittent philosophies on evolution since Aristotle, and more recent dynamic accounts of self-transcendence in the incremental emergence of the ascending holarchy (the hierarchy of wholes). The latter idea incorporates 'lower holons' (from the Greek *holos*,) into more complex holons, as suggested by my compatriot Jan Smuts, Arthur Koestler, and Ken Wilber. The fragmentation of organisms into components is now being reversed; holons incorporate and integrate components into new complex units, more than the sum of their parts, and capable of new selective mechanisms and pressures. However, not only is neo-Darwinism more familiar and mainstream, but its overall familiarity enables easy comparison. I am aware that new refinements, such as epigenetic inheritance, and competition between groups and not individuals, have filled many caveats to the traditional argument, but for the purposes of this book simplicity will suffice.

Since Darwin's *On the Origin of Species*, the broadly accepted view of evolution is that evolutionary advance has been the result of random changes (mutations) in genetic structure, which led to small differences in the characteristics of their carrier, affording advantage or disadvantage in the race for survival. Natural selection determined whether the endowed animal (and through breeding, its species) would survive and reproduce, or would succumb. This theory has been refined, and the detailed structure of DNA has given it credence as a mechanism.

Many things challenge evolution's sufficiency: the inadequacy of the proposed time scales for adaptation; the hugely accelerating speed of change the more complex creatures become; the fact that most mutations are deleterious or confer no advantage; the 'fine-tuning' of structures that would not initially afford any benefit; critical 'quantum jumps' with huge statistical improbabilities; and the qualities most noble in humans that seem to run counter to survival and competition (heroism, self-sacrifice, humour, empathy). There are many others, with new ones coming to light with regularity.

Darwin always acknowledged deficiencies in his account, particularly in the sudden big jumps, and admitted 'Nature is niggard in invention'. Small intra-species changes might satisfy his hypothesis, the survival was plausible, but the new arrivals and the larger leaps were more difficult. The sudden appearance of entirely new kinds of creature altogether seemed inexplicable without new information. It was assumed that gaps in the paleontological record would be filled, reducing large gaps to smaller ones, but that did not happen sufficiently to make simple competition or incremental mutation rapid enough. The pre-Cambrian explosion of different Phyla is a challenge to slow, conservative alteration.

Since then, alternatives to competition and slow continuity have been offered, such as 'punctuated equilibria' with long periods of stasis followed by rapid change; hypotheses of symbiosis, in which cooperation not only accounted for the specialized organelles in the universal eukaryotic cell but mutually beneficial mechanisms permitted the arrival of entirely new forms of life; and the self-regulation of the biosphere.

Schools of neo-Darwinism now vary in adherence to pure mutation or other forms of modified causation. Yet matter governs all of them. Genes and mutation are still seen as the levers of change, even if refinements have been introduced in obedience to the subtlety this perfect, integrated universe requires. Morphic resonance introduces the idea of past habits shaping probabilities that stabilize and create forms according to their old, yet still evolving, 'scriptures', including the evolution of the universe itself.

Mind and matter remain separate even in these recent hypotheses, although the gap is narrowing. Crude competition is no longer seen as an adequate explanation for intricate interdependence and cooperation, or for economy and inventiveness in which the dormant newly revive, the vestigial are retained, the archaic return, the pachyderm persists, and the antique—such as insects and reptiles—are unchanged. Evolution produces spirals, as a spiral produces it. The head of the sunflower would be a better image than the tree of life. But that is the effect of time-bound collective thinking, and we have to begin there, *tant pis*.

The Theory of Involution.

Involution introduces small but critical changes to evolutionary theory. Because the book that follows traces a familiar chronology, the essential contributions made by involution, and the critical difference it makes, could be easily overlooked. By setting out its essential differences as simply as possible, that is less likely, and the journey will be lighter for it.

In involution, cause is no longer accidental mutation but spontaneous act, whether impulsive or deliberate. Behaviour, or act, is the critical driver of change. That act arises spontaneously—in lower forms through the simplicity of structure with positive and negative charges; in higher forms, in response to the pressure for survival. In man, it is the act of thought in response to the pressure for control (through understanding) followed by the manufacture of tools. Information travels in both directions, from and to the genetic blueprint, constantly modifying DNA and its operations. Time influences from both directions too, past and future, but creation happens only in the present. This hypothesis resonates more closely with recent understanding of the epigenetic modifications of the expression of DNA, different genes activated or quiescent through intervention by chemical or mRNA 'switches'. It also puts a different horse before the cart: a mind-driven horse. Anyone familiar with the work of Alfred North Whitehead will find this reminiscent of his understanding that 'Time not space is the key to the relationship (between mind and matter)' and 'Every actuality is a moment of experience … succeeded by a new moment of "now". Experience is always "now" and matter is always "ago" … and one informs the next.' This theory of involution seeks to make Whitehead's drops of experience explicitly creative, and this work presents the evidence through time.

1. **Interaction Leads to Interiorisation** (Involution.)
    Interaction, from the creation of the earliest atoms and molecules, through to unicellular organisms and multi-cellular aggregates, to

diversified plant and animal kingdoms, causes the 'interiorisation' or 'involution' of those positive and negative encounters. The record of oscillation, attraction, and repulsion, is built into the fabric of matter, into its atomic and chemical structure, which reflects a growing complexity of form.

2. **Internal Selection Increasingly Overrides Natural Selection**
Internal selection replaces blind natural selection. The cell or creature spontaneously *acts,* and that *action* affords the advantage or disadvantage. Some kind of action, from rudimentary attraction, repulsion, instinct, deliberation or inspiration, determines the advantage by which natural selection selects. 'Mind', however mechanistic or electrical initially, is the driver of change, becoming increasingly discriminate through this incremental 'involution'.

3. **Interactions Between Organisms and Environment are Retained as Memory**
Interactions are retained in molecular and cellular memory, again leading to increasingly complex autonomous behaviours that range from simple reflexive interactions to instinct, trial and error, insight, and flexible patterns of learning; all of these are earlier forms reintegrated to permit greater adaptability as consciousness rises through interaction with more-complex brains. This increasingly discriminate action leads to acceleration and convergence. The more complex the 'mind', the more autonomous the creature; culminating in the emergence of a single, world-adapted species, *Homo sapiens.*

4 **Matter is In'formed' by Mind**
The development of the brain and central nervous system up through the evolutionary ladder is the response to the interiorisation of encounters that enfold the 'outside' to the 'inside'. This is also involution: memory informing (or in-'forming') matter. In the animal kingdoms before the emergence of man, the consequence of involution is an increasing independence from the environment by the development of behaviours conducive to anticipating and forestalling: social organization, food storage, hibernation, and migration. The anticipation of the future is 'inbuilt' through the experiences of the past. Before mankind, the fixed action patterns

and instincts led to social and environmental integration and are believed to be 'hardwired' or innate. Yet even instincts show a gradation of 'interiorisation' and are able to be adapted and altered, as the domestication of animals and autonomous complex behaviours, such as bird and fish migration, have shown. Animals are increasingly demonstrating that interaction with humanity and his artefacts has increased both adaptation and intelligent problem solving. Interaction and integration continue to govern change throughout.

Man, the most flexible and advanced of creatures, has to learn everything anew. He is endowed with the nervous system and brain that evolution has provided, with their records and abilities, but the capacity to learn anew is what distinguishes him. Man's need for a mind that can learn for survival, by constant interplay with the 'field of consciousness', calls for involution to explain it. Man's dependence on his mind raises a question that is not easily answered by the idea that mind is an incidental consequence of brain. Nor does it explain the asymmetry of the human brain and the way it deals with information: allocating context, wholeness, identity, humour, and paradox dominantly to the right hemisphere; yet analysis, categorization, division, safety, and the drive for certainty to the left. Both have played critical parts in mankind's progress. Involution may well explain this dichotomy in both structure and function, since it proposes that consciousness (mediated by the right brain) unites, while left-brain intellect divides. One led to progress and critical jumps, the other to conservation.

5. **Memory of Evolution is Stored**
Since it is the only chemical structure conserved through time, the most obvious candidate for such memory is the DNA macromolecule, either as the storage molecule or as the receptor, 'reading' the information in the biosphere through shared harmonics or resonance (laid down by shared history), memory of the entire evolutionary experience from the big bang (or its alternative: pre-existing Universes) onwards, in every cell, in every creature, and throughout the biosphere. Consciousness links each to all. Evolution is therefore a collective advance through interaction, not merely simple competition.

6 **Involution in Man Occurs Through the Recovery of Memory**
Man emerges as the single dominant species with the entire memory of his evolutionary path, and its connection with everything else, in his subconscious cellular consciousness. Involution takes a higher turn. Through the act of thinking (contemplation, reverie, dreams, and imagination) and the use of tools and language, mankind re-penetrates this store of knowledge, progressively transferring his insights of connections in his memory to the collective intellectual understanding of his world and his place in it. Through the division of his intellect from his consciousness, he comes to build (collectively) a model of his memory. This is the scientific model of creation. It builds by reversing time, moving from the present back through the past.

The created world seems 'separate' because it has been structured by separation: the I (as eye) observes the 'it'. The oscillation between positive and negative interactions that produce matter also governs thought and the structure of man's intellect, which oscillates either/or, thesis or antithesis. The great moments of synthesis, the both/and eureka moments genius has contributed, are visits made to memory's store of the evolutionary experiences, before words existed to describe and divide them. Science's central concepts—symmetry and conservation—which overarch all theoretical models and are expressed in mathematics, are the collective manifestations of this oscillation between intellect and the underlying consciousness (of which intellect is unaware, although cells and dreams retain).

The chronology of scientific thought reveals that penetration; the process that began in unity (the uniform early universe) diverged in proliferation (the emergence of diversity in all forms of life), and then reapproached unity in man (encompassing all), is recovered in reverse. The recent renewed interest in Greece, in pre-Socratic thought, and in 'primitive' societies and Shamanism, is a symptom of memory's penetration to the very dawn of civilization, and of a hunger for reunification. It seems driven by an urgent impetus towards those societies that retained unity of belief and experience, which characterized early man everywhere. Consciousness has returned to its origins, and neuroscientists, anthropologists, and quantum cosmologists are the symptomatic emissaries of its reach. Yet the presumption of conservation continues to govern hypotheses; the shortage of matter now demands explanations, but those explanations are shaped by the past, equally 'conserving' past understanding. The possibility of an infinite universe of consciousness transgresses conservation, and is not considered.

Early man and early civilizations felt embedded in the whole, and their cultures reflected that sense of participation in all. The impenetrable mystery of creation was embodied in myth, religion, and ritual. However, the incremental recovery of memory, and the control that understanding afforded though the invention of tools, led to the separation of intellect from (un)consciousness and the externalisation of matter—its separation from mind. The divide between mind and matter (the Exile from Eden) has been the result.

The perennial philosophy of mystical truth and esoteric disciplines retained those original mysteries, but expressed them differently depending on the cultural demands of the time. Mystics in recent periods, such as Teilhard de Chardin and Sri Aurobindo, see spirituality within creation, not removed from it. Mystics have always led from the future, holding up the torches of direction, but in their languages and integration they too demonstrate the influence of involution and the recovery of the whole, the spiritualization of creation, rather than spirit removed, or 'above'.

The role of involution in underpinning the choices made by consciousness restores the return of mind to matter, as matter-mind, and mankind to the whole of consciousness (or the self-created God).

Whether involution will be acknowledged as a theory or dismissed as an unsubstantiated hypothesis will depend upon the nourishment found in this new odyssey and on any marriages it might make with other minds that approach from other directions.

What follows?

This scientific hypothesis echoes every mystic leader or religious tradition. It reintroduces the Buddhist 'God', consciousness, as acting throughout creation. In tandem with the emergence of creation in rational man, it co-creates a 'self-knowing' God with all his traditional qualities, immanent, omniscient, and undivided. As a theory, it has been anticipated by Père

Teilhard de Chardin's 'noosphere', David Bohm's 'implicate order', and Carl Jung's 'collective unconscious', and recently restated in Ervin Laszlo's 'akashic field' and Rupert Sheldrake's nested 'morphic fields'.

The concept of a seamless pervasive field of consciousness manifesting in the material universe also makes room for multiverse theory. Our universe may be merely one of many, or a succession of many. The informing intelligence embodied in the process that produced mankind may well have refined its trials in other universes that preceded or coexist with our own. The limits time imposes on evolution are eradicated through involution. Greek philosophers' early intuitions of the mathematical order that underpins creation, and the patterns in number and harmonies repeated throughout, would logically imply the pre-existence of such an order, which was accessible to early, undivided intelligence. Our cycle of involution may simply be consciousness navigating a new spiral, with new forms and new creation.

With regard to evolutionary theory, involution explains many things random mutation cannot, including acceleration, perfection, mutual interdependence, sudden emergence of synchronous species, and the improbabilities of critical steps in evolution. Impulse and inspiration have no causes; they may have timing. Timing might be the product of a collective pressure for change, which, like the pressure of artesian water, finds an outlet in the ready mind (including the instinctive minds of animals). It may also have direction, since convergence towards unity has seemingly been the pattern leading to human beings—as yet incomplete beings that are unaware of their unifying spiritual potential.

Involution offers collective science an opportunity to re-examine its methods and conclusions, and offers man an opportunity to build a better collective 'model'—a model that recognizes his dominion has limits unless he integrates with consciousness and learns to live with a recovered God, the perfect creation that has been built through the consent of all to each.

It makes yogic and spiritual exploration—realms of consciousness hitherto ignored as the province of cults and crackpots—the new direction for science itself. Matter is merely a particular 'state' of consciousness, almost illusory (Plato) when viewed from another state (Einstein). Relativity may describe states and interactions of consciousness. Involution reconciles the thorny difficulties faced by quantum mechanics and relativity, the multi-worlds posited by string theory are perhaps reflections of science's awareness of states of consciousness, enfolded and obscured within

Introduction

the only state science recognizes: the material. Their invisibility may be no more than a product of science's blindness, the opacity of intellect that covers the eyes of thought.

Involution also explains many things regarded as bizarre or fanciful: memory of past lives; the 'finger memory' reported by musicians, which cannot be explained in terms of brain and nerve speed; déjà vu; premonitions; the simultaneity of new events widely separated in space like the growing of new crystals. It explains events requiring the speed of thought, not light, and all the inexplicable entangled quantum events as manifestations of consciousness, not of matter. Telepathy, clairvoyance, and near-death experiences are all travels in the realm of consciousness, well attested but inaccessible to measurement. If miracles are the subservience of matter to consciousness, then, through involution, they have been with us always in the miracle of creation.

So Why Poetry?

Evolution as a concept has now broadened beyond the realm described by Darwin—the world of biological diversity—to encompass the origins of creation, the start of time, the life and death of stars, the appearance and disappearance of matter in black holes, and quasars or the Higgs boson. Science now stands astride the gulf between relativity and quantum theory, the last of the seeming irreconcilables. The continuum of all scientific disciplines is encompassed by an evolutionary explanation without a prime mover, a God, intelligence, or purpose. The collective intellect of science has constructed a single, unidirectional flow of creation.

Science, as practiced by the collective, is a left-brain affair—a slicing, hierarchical, categorizing store cupboard—although many of its creative giants were perhaps led by the creative 'master" of the right brain. These imaginative, context-conscious, idealistic, relationship-seeking, flexible, creative innocents re-asked the questions that others believed were already answered. Involution as a theory now seeks, through poetry, to readdress the master of the right brain and all of the explosive leaps forward made by its creative genius. It was through genius that a newly perceived 'whole' was handed to the industrious left brain, to analyse, dissect and corroborate.

Unlike music's evocative appeal directly to the connections 'between' (between keys and emotions, between notes and silence, between suspensions and resolutions), the structure of language—its meanings and definitions, its dependence upon known patterns and concepts—is the closest manifestation of left-hemisphere, partial 'either/or' logic and the explicit. The meanings of words, and the grammar of time and sequence, all depend upon left-hemispheric processes: building on the past, and comparing and evaluating according to pre-existing patterns.

The metaphoric language of poetry, which is closer to music, deals with the implicit, the paradoxical 'both/and' that offers its revelations to the right hemisphere. And it is in the right hemisphere that the whole—the context, the allusive and intrinsic—finds purchase. Poetry has always been the language of the mystic, and involution is mystical science.

We have, through collective involution, now recovered the all, and we return to the whole of the uniform beginning; the cosmos needs the computer to imitate the mind that understands it and to utilise the integration the mind fails to recognise. To retrace the history of science with the accompanying evidence would be not only unwieldy, but indigestible. Besides, it is too familiar.

Poetry blurs the trees for the sake of the wood. The painter Seurat conveys a river bank on a Sunday through small daubs of colour, and since the viewer already knows what Seurat alludes to, the detail is superfluous. That is my hope: by taking a blurred travel through the history of science, the chronology and sequences that testify to involution will become apparent, not just the facts themselves. I hope to have judged how many significant ideas were needed at any point to make a continuous rope by which to hand evolution securely over to involution. Always, I have held the collective in mind, yet the knots were made by the individual geniuses, with whom I sojourn long enough to give the essence of the difference they made and the areas they illuminated. Many will fault my omissions, and to them and their favourites, I apologise.

If the history of man's intellectual understanding is the steady, incremental recovery of memory (and his reapproach to consciousness), then involution should be equally (and more easily) shown through other effects: man's art and music, architecture, cuisine, and literature. I believe it is. From the simple, flat perspectives of cave art (or the rhythmic simplicity of music with a repeated beat or a single voice) through to the increasing complexity of depth (perspective, harmony) and the greatest complexities of classical composition (sonata and symphonic form, or classicism's perspectives) back towards dissolution (impressionism, cubism, abstraction or atonality, and the bar-less beat), involution reveals the underlying fluidity of energy and reapproaches the start of time.

Introduction

Man's creative mind produces his instinctual sense of his stage in the recovery of his evolutionary path, from simple (unconsciousness) to complex (divisions and structures), and back to simple (uniformity of matter, infinity of mind).

Poetry attempts to approximate that, enabling the freedom to allude, to evoke in broad images the wash of consciousness through the human mind, from the present to human origins. Yet I am aware of poetry's limits, and of its dangers.

In our modern era, Wittgenstein echoed Parmenides in suggesting that the proper language of philosophy was poetry. Philosophy is essentially a right-hemispheric realm of thought—intuited not deduced, all-encompassing, seeking connections, drawing forth deeper insights, and binding together. To use the limitations of the left hemisphere's grammatical prose to attempt to convey it is to imprison the bird caught on the wing in a cage, denying it flight.

In writing the theory of involution as poetry, I face the opposite danger. In attempting to score this poetic account for the imaginative, musical right hemisphere, I have only the left hemisphere's collection of categories—the chronological, didactic, hierarchical facts and lists of names. Poetry cannot nail them to the oak; only fleetingly chalk their rapid passage. The danger is that what I hope will be conveyed as a reliably evidenced alternative to Darwinism, remains insufficiently explicit or trusted to bring about the new vision involution might offer. The right brain may appreciate it, but make no mistake- the left is still in charge of science, of the world, of economics and of respectable theory.

Through poetry, the bird of theory might fly, but scattering its feathers and disappearing from view. It may not linger long enough to be seen at all. In the hope of controlling this danger, I have accepted many constraints in poetic liberty—some unfashionable narrow traces, some rhythm, some rhyme, and alliteration when it found an easy perch. In this era of formless art, any structure seems to bring a pursing of the critic's lips; even music lacks a beat (which I permit at times, hopefully discreetly). Rhythm is one element of music to which the left hemisphere responds, and I want to keep it on side if possible, but not at the expense of giving it too much importance. This poetry is to be read at a gallop, enjoyed for the pace, and not the chafe. What it evokes will ride easily, what falls will not matter. The footnotes provided take no knowledge for granted and may flesh out allusion or be ignored entirely. They are sketches in a notebook on the move, neither definitive nor comprehensive. Others, like Wilber, offer that exhaustively.

There are further problems in trying to discuss a theory using poetry: matters of aesthetics and liberty. I could strangle poetry at birth by summarising what each canto seeks to convey, thereby rendering the reading superfluous. I could summarise each canto after you have read it, thereby amputating the post-cadence reflections (as anyone at a concert knows, opinions from the adjacent seat kill the music's magic). Or I could try to ensure that this theory of involution (the Return from Exile) is spotted like spoor as the journey progresses, by placing 'pace Darwin' markers that alert you to the small but critical changes it makes to current evolutionary theory, which seems the least ponderous option. Footnotes relevant to involution, and important contributory facts, will be numbered in bold typeface; (and Involution is sometimes capitalised for emphasis, for it is fairly soft-spoken): others merely expand poetic allusion or references. Both wait to be called upon if required. Those towards the end are fairly technical, but simply summarised, only because no knowledge is taken for granted. Further, headings indicate the gist. Because I do not assume that a reader will read all of the footnotes, there is occasional repetition of information. The same supportive facts may be needed at different times, in support of different points.

This broad-brush overview of the rough terrain of the theory of involution has hopefully alerted you that, throughout history, consciousness's patterns are recognized by its distinctive and intermittent traces.

* *The Master and His Emissary: The Divided Brain and the Making of the Western World.* Iain McGilchrist's superlative analysis of asymmetric hemispheric brain functions and their respective influences on cultural epochs. (Yale University Press, 2009)

Canto the First: The Map and Older Tracks

Does Messiah part the cloud when man his origin forgets? [1]
In another epicycle, another set of clothes,
To speak as is appropriate in the play upon the page?
Rather as a comet comes, in periodic phase.

*

Siddhartha lived on lotus shoots with his enigmatic smile [2]
Sat instructing devotees with one collarbone exposed,
Trailing bowls through dusty streets, garlanded with flowers,
His music Glockenspiels of wind through hanging pipes.

He departed softly in the fullness of old age,
Leaving temples with tip-tilted eaves still wearing his sourire.
His monasteries are filled with doves, quiet, sandaled men
With shaven heads, modest means, naked well-tuned ears.

*

Next incarnation had him stretched between divine and desolate, [3]
Turning over tables, feeding multitudes with fish, preaching pithy parables…
Women wept and washed his feet; friends counted silver coin,
Increased the talents, as advised: financed the Vatican. [4]

He donated agony, eternal grief, wounds and vales of tears,
While soldier priests cast lotteries for copes and influence.
He is remembered most for jam tomorrow…
A scrape of dripping for now.

*

Involution

Perhaps upon his next traverse he reflected that smooth pearls [5]
Were wasted on Sodom's cloven-footed swine;
He'd suck on pomegranate seeds, enjoy his several wives …
He'd spell it out…Make no mistake. He'd dictate the Qur'an. [6]

This time He would leave no doubt that God had cloaked his face,
Abandoned vain attempts to give man any noble choice…
Not worth the candle of one suffering son.
Instead He tossed a brick-like book through the garden door…

Eden was game-over. He would not return.

*

'Forgive me this intrusion, but is this relevant?
I had understood we intended to dissect
The arguments of Darwin and others of his ilk,
Not sail in junk conjectures, the broad estuaries of faith…
Can we narrow down and tread the paths towards
A modest goal? A small thatched adobe town
With a pillow and a meal?'

In tandem with appearances of Teachers in all climes
In times past and periodical, there were more loquacious poets
Taking odysseys of varied length, in motley company, [7]
Perhaps they too were single Soul, diverse tongued,
With an appetite for travel.

Homer may have been one man or a band of bardic brothers [8]
Sending up the chariot dust through Agamemnon's wars,
Which had long settled, leaving no traces except
Medals for Hector's courageous buckled breast;
Achilleus threw tantrums: Languid Paris swooned. [9]

All of human frailty was the butt of song and the Gods' amusement
Until man came of age. It was the blind poet's pleasure
To sketch a plan of the polis; prepare boulevards of marble
For Socrates' hemp tunic…blood the battlements of lusty Troy
With its static nursery horse.[10]

Canto the First: The Map and Older Tracks

This first journey linked Olympus with Mycenae, all of Gods and men[11]
Were the playthings of their passions, the hot headed and the calm;
The impetuous in combat, the cowardly in tents…
'Be careful what you wish for'… Until the Gods were mollified
By slitting something's throat.

Yes, yes, we've heard it all before,
This diversion ails…Can we cut to the chase and expedite
*What **this** narrative intends?*
Why drive through prehistoric sites… unless the cogent evidence
Is buried in their mud?

Journeys are the substance, their patterned poetry
Take Gods and men and intertwine them like two convolvuli,
Of symbiotic spirals, needing one another's weight…
Poet too may reappear, on nightingale repeat… [12]
Recovering old rhapsody for newly alien corn.

By the time the Tuscan set out with his Roman guide[13]
Who had consigned his hive to winter's sleep, his vines all neatly tied…[14]
The intention was to mundi mappa all degrees of sin,
On Hades' cold escarpments, before trudging through repentance
To paradise encircled by Beatrice's soft arm.

Womankind the beckoning, the portal to divine: man in need of softening
His insistent song; of self, of work, and trivial things
That keep incipient tears of joy from spilling down his cheeks.
Each caravanserai of travel seemed only to repeat
The fading unity of One, the Many and the All.

Each succeeding bard had a different bag of tricks;
Homer scratched his attic song on Olympus' slate (so close were men and Gods)
Dante slipped out in velvet shoes to seek converse with the damned
Who remembered vice's pleasures they had learned in wayward lives,
Their menagerie vernacular, now terza rima rhymed. [15]

Involution

There were others later: Milton, Goethe spring to mind,
Paradise lost and Paradiso found... leering Mephistopheles [16]
Tempted avid Faust to make checker play with intellect...
While Gretchen patiently ribboned her braids
Dangling the golden keys to heaven.

In early nudge reminders the Gods were everywhere:
Grey-eyed Athene dropped in to dine, Hera waved white arms...
Displeasure was appeased by barley and fresh blood,
Last suppers were convivial; black ships departed ports,
So pawns could serve divine ennui in antique quidditch sports. [17]

After Aristotle the mountain carved with terraces instead [18]
(The ascent in granite narrowly defined)
Keeping each man an allotted place a certain distance from God.
Hierarchies of virtue applied as much to men
As to impervious planets orbiting the sun. [19]

Always the same story, the forgetfulness of God...
The poet saddles up again and leads the horses out...
If I wish to mount and trace a path, what metaphors remain?
The new religion pickled and in a test-tube shrivelled.
A vocabulary of brittle tones... How may I coax them back?

'Ah, at last I see wither we have travelled
Via all the others that saddled up a mule.
(Whether emissary Son or devoted poet)
Are you calling the weavers at this loom
One single and the same? Obeying the same summons...
Spaced in intervals of time?

Perhaps I am.

I know the true poet finds his words
In charcoal dreams, a soft-shuffle roundelay
That emanates from reflections in pools of solitude,
While the tongue licks a pencil, to hear the heart's pause
Seeking the metre to echo, the right line to lead.

Canto The First: The Map and Older Tracks

Mostly he wraps silence in a paper screw of words
That offers the liberty to linger on the grass
Where clouds above scud images into personal tastes,
Tonguing newly minted thoughts, striped zebra sharp
With wintergreen, storm liquorice, or dripping caramel.

To take a graphite pencil to the business of a quill
Confuses the pentameter with the pterodactyl… [20]
To murder liberty, forsooth, it is poetic crime!
But argument will feed on poetry's fresh corpse…
Virgil detailed a slaughtered calf to revive an ailing hive. [21]

Bees are not irrelevant. The messengers of Gods
Replenish stores for winter, by singing summer's song;
Bees are every flower's health, every hedgerow's reason…
Now they are a-dying, a drowsy fatigue falls
On industry. The world is sick and takes the bees to pay.

For hubris, for arrogance primping in its tower
Impervious to troubadours plucking lutes below…
The casement window slammed upon a dulcet song…
The hungry hope silenced that belief might invite
Spirit to attend the feast if signalled to come up.

'*You were doing much better when you couched with bees
Let's avoid the rant. Use your quill to lethal purpose,
Give owl-wise argument.*'

I do but clear the site to build: I start with sterile bees [22]
With no happy conjugation, they build soft cells of wax
For their queen's succession, her multitude of drones…
If she dies, worker turns looter, and destroys what it has built…
This fact should rap the knuckle of the certain Darwinist

Who elevates the selfish gene to commander-in-chief [23]
Like a Hitler edict, bestrewing copies of itself,
Successions of Aryan blondes, with mare-wide healthy hips…
Kinder, Kirche und Kuchen, the permitted variants? [24]
How did sterility find its place through time's slow increments?

Involution

There's more. The language of the waggle dance, whereby returning bees[25]
(In the whirring glimmer) tell the audience that gathers
The angle of the sun to keep, just how far to seek,
In which direction lavender or clover can be found… [26]
Body language in the dark? (Like all those clever quarks) [27]

Bees talk to one another, and work collectively
In colonies of husbandmen, to build, protect and feed,
Clean and repair, seek pastures new, lug the pregnant weight
Of eggs for future dynasties they will never see.
For royalty, if threatened, they lay down their lives.

Now here is a clear example of what I've come to call
'Argument hatched from incredulity'
You marvel at the worker bee
Unnecessarily.
The dance you describe was not always in the dark
But perfected on the surface of tree hanging hives…
While the sun was visible…
It's but a small adjustment to the vertical gloom
Once man had mastered the keeping of combs.

The gene is ingenious and can easily code
Nijinsky's antecedents and the corps de Ballets Russes…
Break down the high lifts and the graceful pas-de-deux
Into components with entrechat, glissando,
The plié, lift, and slide…
Nothing much to it, over millions of years.

With regard to sterility, widen your view
To the hive as single organism like a platoon;
It has subalterns and couriers, signalmen and cook,
Sergeants with whiskers, captains with a taste
For shouting and a whisky in the officer's mess.
Each component subordinate to the task assigned—
To bring home oil and tribute, and try and stay alive…
The only difference is the CO. In the army he's visible:
In the cell, mute.
Each division of labour refined over time. (If sex went awol

Canto The First: The Map and Older Tracks

On labourer's pay, the eunuch worked harder for the sleeping harem)

The gene is born selfish with a single desire
Like Narcissus to see itself in every liquid mirror. [28]
Whatever it took to climb the ascent
Of competitive advantage would be its device:
In bees and termites, wasps and certain ants,
It settled for sterility, assigning egg production
To an efficient machine.
The kibbutzim in Israel tried to emulate the hive… [29]
Divisions of labour would make the desert bloom…
Defend the encampments with ardent fighting young.

What ho! Who speaks? That dull pontificate?
Did I catch a glimpse of an academic gown?
The square-jawed Professor for whom a chair was carved
Better to translate the subtleties of science [30]
For the pitiful believers in any kind of God?
The man who fashioned from a parchment (curlicue embossed
With a laudatum in Latin outlining his skill…)
A club of stone to beat out brains
Harbouring the virus of illogical belief?

I did but represent him, (he would not approve
What he would **not** call poetry of a pupillary kind [31]
But a mote in the eye which hides fallacy's beam)
I did but baldly sketch what he has made his own
(A life devoted to illustrating tomes, tracing
The gene that snakes its way through all the countless
Beads; Hail Marys, Mea Culpas that compose the rosary
Of creation now full stripped of any majesty…)
Threaded now on dog eat dog, devoid of destiny.

Those stanzas just suggest the thin gruel of bones
In the stew-pot that must feed the waiting multitudes…
Camped upon the sun-soaked hill, used to fish and loaves
With wine to toast the nuptial pair… Instead now offered chance
To explain creation, how they came to be…
Why they dream such lucid dreams, why they weep or laugh…

Involution

Their children must use callipers if to this piper they would dance.
We have only a blind watchmaker, tinkering with cogs
Of error's navigation through the needle of the eye. [32]

The gene is not confined to influencing cells,
But spreads, 'extends its phenotype' to construct a beaver's dam, [33]
Provides hosts for parasitic worms, the imitative plant,
The camouflage that conceals an insect, moth or frog…
All are added to the broth to thicken resolve
Against any hint of purpose. If you slice the process wafer thin
Or crush it very fine… A spoonful of soft syllables
Helps all sophistry slip down.

The aftertaste is bitter. How did the intricate eye
With a lens tuned to focus on a curved retina,
An aperture to shield out too blinding a light,
Its parallax corrected by a split visual field… [34]
(With cross-my-heart neurons…on their path to the brain)
Ever trial and error it, or slowly refine
The rough edge to assembly in the initial design?

Einstein would not swallow it: the roll of God's dice, [35]
Or simians achieving 'To be or not to be' [36]
The quantum explanations
Of mathematical sight, were all well in theory
But never enough.

Does that mean in essence you cannot uphold
Darwin's conclusions? In no ways at all?

I accept his bumbling errors but they cannot explain
Most of the things that maketh the man.
His valour and sacrifice, even of life, [37]
His sense of the ridiculous that laughs away sex,
His tears of compassion, with their plumbing of ducts, [38]
Rendering him helpless for fighting or flight…
But mostly his belief, his conviction deep dyed
That something, or someone, cares about him.

It survives in deep forests, for the barefoot Indian
Who calls upon the serpent, (the archaic scaly sage) [39]
To tell how it works, the heart and the mind
Of creation itself, and finds he may,
Through being attentive, cure his feverish son;
Paralyse a monkey, preserve tomorrow's fish
Ask for honeycomb politely with a torch of harmless smoke.

Those mentioned messiahs all rekindled the flame
Whose wick needed trimming, whose lamp had run dry… [40]
If natural selection had four spans to make an eye,
Improve it for an octopus, or two falcon foveae, [41]
To perfect the bee orchid, the incubating wasp…
By weeding out the unfit, so the fittest remain
Why does belief wax ever stronger and conviction multiply?

Nowhere have I heard it said that man is exempt
From Darwinian selection, appropriate fit
For purpose defined by competitive strife…
What possible value in preserved blind belief?

Or wonder? They wave away the questions of incredulity
With another throw conjecture to cover the stain:
The improbably unlikely, the highly complex,
Things too intricate, too perfect for chance
Or time's slow stumbling gait, making more
Detrimental errors than advantageous ones…

Instead of facing wonder full, as the light of a new moon
(That may draw up the skirt of the tidal unseen;
Such lunacy could light a path never yet trod…) Instead
They slice time for explanation… and 'Time' never buttered bread.

Only his gaunt shadow playing chess upon the sands, [42]
He props his scythe against the rock, and enjoys the artifice
The foreplay, the suspension of the certain terminus…
The victory is not in doubt so he'll share another glass.

Involution

Time is an archive of the partially viewed
From over the shoulder, the over-and-done.
The sleeping side of the unchanging past
(A Dover sole, all swivel eyed, flat against the mud)
From half of mind only, the divided mind,
'Either this is me or it. That is matter.
I am mind'.

What? Did I hear aright?
You think you are mind?

When viewing the material I'm compelled to dance in mind
Like those pirouetting couples confined in an orb,
One is stepping left, the other stepping right… (rotation an illusion)
Either this, or maybe that, but very seldom both.
(The orb of glass is the camera brain
Firing per second along neuronal paths…)

No accounting for Mozart
Or the Warwickshire bard,
Beethoven, Einstein, Faraday, Gauss,
Who found colours in music, numbers in space:
The tattoos of immersion in something called truth,
An aroma so precious it eluded a name
All those cowries of comfort for the sea-hungry ear…

Tell me again that the structure of brain
Explains the creations of inspired men?
Or answers any questions as to how a blind mole
Treads the pinnacles of virgin snow in which
Men leave the footprints of God.

*

Are the original eurekas all you have to lob
Against the bastions of certainty? There's a brave girl!
None since Archimedes sitting in his bath,
Inspecting filigree, perhaps with silver sullied, [43]
Could quite give account from where

Canto The First: The Map and Older Tracks

Aerated water suddenly bubbled…
To crown cogitation and leave in its flow
Certainty pellucid, of something not grasped.

The eurekas may set the seal on my hypothesis
(The watermark in paper held up to scrutiny)
Before I bite upon the gold that backed the currency ⁴⁴
(Those immortal thinkers with knuckles on their brows
A la Rodin or Michelangelo, for the latter was one), ⁴⁵
I have a snake and ladder game, a single molecule
Threading beaded information from each and every choice
From the alleged quantum start of time, to the positing of strings. ⁴⁶

This same double helix collected it all.

From the (waterless) farm of cosmic soup ⁴⁷
Where electrons orbit around clutches of quarks,
Heavy lidded neutrons seldom stay awake… ⁴⁸
Where protons and neutrinos lead a merry dance
In and out of theorems, tied to equation's gate
Herding wayward energy from ducking under wire… ⁴⁹
Trapping black holes under upturned pails, while
Cheshire cat infinities grin and duplicate. ⁵⁰

The needle in this haystack now is being sought
In a 'lets have collisions' shaft deep in the womb of earth; ⁵¹
The dragon jaws of magnet teeth all primed to pounce
On ephemeral traces, scratches on a screen. Come
Let us now detect where Higgs holiness resides…
In the interface of hatching straw matter out of mind…
'One field theory with coleslaw, two large bosons with fries
And a Nobel Prize strawberry shake coming up'.

Nobody considers that the cloud of ideas
Like wonder and conviction might have had a role, and more;
Might have hatched virtual mind spores into actualité
By molecular storage, that leaned to predicate,
Packing mind into matter like babushkas enclosed ⁵²
One inside the other, forever with child.

Involution

We calculate relationships with everything and time
Since we parted from orangutan or the forest fern.

Let me be quite clear. What you seem to be proposing
Is consciousness in control of speed and advance
From simple to complex.
Is that the nub of your quarrel
With Darwin et al?

Natural selection is an undiscerning eye
To choose between this 'fitter one' rather than 'that'…
How does 'quality control' somehow select
From creation's widgets passing below, what constitutes the fitter?
When it is 'blind' purpose, and none of them act?

Sure, selection it is, that makes the sudden choice
To freeze in paralyzing fear or make a speedy scuttle,
To pounce on potential prey, discover new tastes…
Behaviour is the outcome of experimental thought.

Impulsive thought, through action, pulled the lumbering cart
Of evolution's progress uphill against the slope
Of chaos and inertia both drugged by entropy…
Instead took quantum leaps up higher narrow crags, [53]
Unerring, sure footed, on four cloven or two.
The creature chose before it was chosen;
(The hare before the tortoise that followed in its wake) [54]

I am not suggesting that thought was waving any wand,
Or, in its cosmic infancy, was beyond electric charge [55]
That selected an appropriate element bond
To join the Ouroboros of Kekulé's band: [56]
To whirl organic circles in a mocking benzene ring.
Linkages grow complex from simple elements,
Each twist of the helix more than sum of parts—
It takes five synchronised mutations to reverse a single step. [57]

The great river of creation that gathered tributaries
From aeons spent in trickles or the Brownian dance

Canto The First: The Map and Older Tracks

Was railroaded by Darwin as a message for his age,
The discipline of ruling banks was clearly understood...
Random incremental change given six of the best
Bare buttocked across the Headmaster's desk.
The race to the swift, honour to the strong
The length of a wicket had been branded in his brow. [58]

This very day, I propose to explain how it came to pass
That mind was severed from its fertile conscious self,
Through its own cogitations and opposable thumbs:
Peeling fruit, using tools, making wheels and clocks
Lenses for the heavenly spheres, ploughs and gyroscopes
To navigate the oceans, subdue capricious seasons
With an irrigated Nile. With increasing mastery
The distance from 'it' grew. 'It' was subject, outside itself. [59]

By the time Darwin sowed his cautious conjecture
In the ploughed furrows of the carving intellect
It saw the world external, quiescent and subject,
Ripe for understanding, in vectors and graphs,
Acted on, not acting, except to Newton's laws.
Man was now omniscient, but never thought to ask
How came his mathematics to so perfectly fit
Everything that preceded the arrival of him?

Mind, the unique, superlative crown
Had somehow happened, with the burgeoning brain
In the last two minutes of evolutionary time. [60]
An epi-occurrence, another accident
By-product of complexity and natural dominance
Over lesser creatures without intelligence...
They had instincts hard wired through competitive strife.
(Since speech is never evident, they must lack consciousness)

That Self, so coiled below the strata of mind
With its web of connection to every living thing
Struggles up in haunting dreams in tossing restless sleep...
Fills poet's heads with fears of eating a peach... [61]
Runs seductive fingers through imagination's hair...

Involution

Maddens painters in Provencal sun into cutting off an ear,
Scribbles semi-quavers, cascading for the deaf
To callus the fingers of a fiddler and leave a bruise upon his chin.

The Self is mute until addressed
By a longing deep pining for a beaker of the South... [62]
(Dear Heaven... please, a purple stained mouth)
When Pythagoras strolled past a ringing forge
Above the Aegean, in its lucent first light,
Mathematics and music were together tightly spliced
By his perambulation and the pitches of sound
Of octave span or perfect fifth, the second and third, [63]
Nothing is simple in the simple monochord.

Music shaped by what's not there, the same notes play
A reverberating organ fugue, a simple canon
On strings, through reed, in percussion or brass...
The silences between call forth the melody...
Even so, below the riverbanks of chance and survival
Coils the undulating river scaling for the sea:
Michelangelo's creation with its newly parted hands... Do they
Crave equally to clasp again to unite the severed self?

Darwin forgot the silence in the music. Heard only separate notes
Vying for precedence, competing to be heard...
Harmony subordinate, clashing swords and shields
Of competition, strife, and hard cacophony.
Science so soaked in one against another
Thus genes are 'selfish', 'moderators', 'outlaws on the hill', [64]
In the 'arms race' of advantages if they co-operate
They only do it to deceive, to feint and so attack.

This is the language of divided mind, everything opposed:
The either-or is me not you, my kith but not your kin.
Yet we share most of what we are, in that most careful coil
Recording time from its beginning and which permeates
The entire atmosphere. In space, in stars, in interstellar dust
Viruses queue to ride the cosmic dipper...
Bacteria book the meteor jaunt

To re-infect our gene pool, with newer variants. ⁶⁵

All speaks to all, as grass when broadcast
Appears first in green bubbles where the hedge gives shade,
Joined later by struggling drier seed, it makes the sward
Full velvet by joining fescue hands.
New crystals grown in Montreal create copies in Peru…
Once born, existence is assured elsewhere and far away…
A dragonfly may initiate monsoon in Kathmandu
By struggling in Tennessee to open iridescent wings.

Each major ascent up the convergent cone
Gathered a greater need of the whole;
Stitching the sampler's alphabet
In four tone codons spoken by cells ⁶⁶
(At each intersection the possible dulled
By collective refusal to anything new)
Brave new ideas are sensitive
To antigen attack from the body politic.

You would turn the argument through a semicircle
To face the other way? Instead of bumbling accidents
You place consciousness as charioteer guiding the reins,
Soft handed on the horse's mouth, swift as falcon flight?
The unerring path, the certain ride through the desert dust,
The foam flecked mouth, the glistening flanks, bloodied by the lash
Of increasing haste to sink a muzzle
In the flowing cold river of God?

*

Would you go further and assert
That God waited upon the ascent of man?
To discover His nature from unconsciousness
Becoming through slow evolution aware?
Creation and Creator hand in hand
To find themselves in complement.
Lifting the omega diadem seal
On creation's sole and purposeful trek;

Involution

Man married to All, and God become Man?

With caveats and much more time
For man to explore his deeper self
In unity with all that is…
Which could begin with modesty
Spent examining my postulates…

I would.

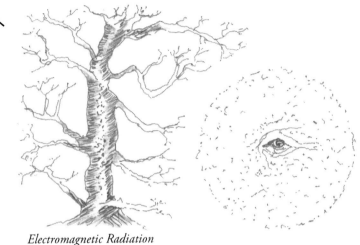

Electromagnetic Radiation
THE TREE OF LIFE   THE EXTERNAL UNIVERSE
Divergent: Equivalent   *(Expanding: Steady State?)*

EVOLUTION

INVOLUTION

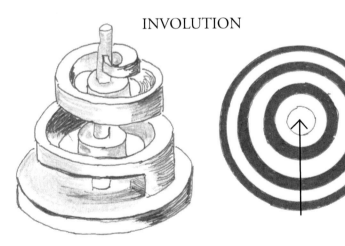

THE CONE OF CONSCIOUSNESS   THE INTERNAL UNIVERSE
(Increasing Complexity Convergent)

FIG. 1

Canto the Second: Memory, Dreams and Separation

Since we are now together underway, the axle sways…
The beasts begin to doubt the dimensions to traverse…
(A journey through all thought about
Creation's patterns and its processes)
Hitherto so neatly interleaved
In covers of crisp discipline
Defining small legitimacy…but
Have you not already lamed your foot
By drawing your conclusions at the start?

It is the habit of the scientific mind
To give the 'abstract' to a ragged boy
Who leads the caravan through unmarked sand,
Whistling to himself in hope
That plodding evidence will follow in due course. [67]
I will unpack the argument in parts and trace
The larger liquid circles spread
Like wavelets from the stone that fell
Pom…falugo…per…phlas…matosh…sh…sh…sh… [68]
The onomatopoeic plunge…the heaving swell…
The dissipation at the edge…
Slowly the water restores the sky…
Till all is still.

So creation returns to its initial state;
All energy is one, (a changing gradient)
Turbulence local; rings of smoke…

Involution

I would remind you of your mission yet,
The wheel of knowledge to be turned full round.
The inertia you would seek to shift
(The weight of all collective mind)
Has embedded it full axle-deep
In certainties.
Take it bit by bit
Transparently, as lucid as
A small Bach suite
And slowly set
A steady walk
Towards an alternate.

The music of the word must like a thumb
Erase emphatic lines; suggest the shadow of what lies behind
All claims, or else the dervishes will contradict,
Whirling in pointless ecstasy, to prove you wrong.
Much have we travelled through the world of dream;
The membrane separating day from night
Thinned by your frequent forays through
From thought mundane to wider views…
That interface now must improvise
A barge of words, caulked, water-tight
To ferry understanding.
No trampoline for intermittent height,
That subterranean, inchoate stream
Must steady at the surface flow.

The rational intellect remains confined
By subdivision, on a precipice
Of looking down upon creation's state
Far, far below, although in truth
We are surrounded and encapsulate.

How can I escape the tendril traps
Of words' accumulated meanings, not my own?
Conscript those porters for untested paths
To wend an argument through un-trod grass

Canto the Second: Memory, Dreams and Separation

And have them follow willingly?

Use words to summon pictures to take shape,
Like Keats's attic urn; spare, shadowed, still,
The parade of cattle, garlands, maidens, men,
Preserved in aspic time, upon a shelf
To catch the light but through the poet's pen.

Imagine such a shape, a simple gourd
(Such as might hold water on an Afric head)
A Greek amphora, narrow near the base
With swollen hips towards the waist,
Contracting sudden to the open throat…

For which contrive a handsome cork.
Burnt with the vintage of new vision, raw
(It may, perforce, in years, mature)

That is the whole creation now entombed…
Slow to expand through aeons at the start…
The Cambrian explosion: All at once
Prolific phyla synchronous appeared…
Abrupt and sudden it contracts,

Closes off, begins, the argument at Man.

That perfect pot more apposite than tree
To encompass all creation's shape…
Appearing motley only from above,
(Equivalent all forms, to the myopic eye) Stand outside:
Convergence will ferment a single malt. [69]

This gourd contains the all, and Man
Has gathered up the all within himself.
His cells, his structures, chemistry, collect

Involution

Full tribute from the path experiment
To perfect the human body and its hands

That with his mind must now explore the rest.

Some themes, perhaps, to set an overture:
Intelligence as pathfinder, moccasin through shade…
The separation of the mind from deeper Self…
Opposition in the ladder rope ascent
To newly knot the simple in complex…
Acceleration (as increasing consequence of thought)   70
To single sapiens, forgetful of the rest.

Have done with generality,
Give the meat.
Tough argument requires teeth,
Tear off the gristle of apology,
On a salad board serve up
A fresh outline, sufficient to itself.

Start with random error;
(The usurper of creation's pulse)
Expose the sham of its pretence
To rule more than the kingdom of small coin;
The counting house of small exchange
By which a species satisfies itself.

Start where we stand, then move through
From what is known, to pastures new.

Very well, I'll do my best…

It is believed that error in a molecule
Was time's apprentice, and sous-chef
Of changing dishes, served on nature's plate.

The wide expansion of the tree of life,
With all its limbs of slow ascent, diverse,

Canto the Second: Memory, Dreams and Separation

Made not a single change in time's reverse…
Time is the underlying minuet:
Its signature (written left to right) mistakes
One single cause to come before event.

The creature at the whim of the minute
(The microscopic diktat from the cell)
Produced a synergy, a mesh throughout,
Intimate and self-correcting, but
Instructions had, themselves, no feedback loop.

Did not amend the recipe itself…
Improve the planning, or economise
Through closing off the cul-de-sacs. But here

Time moves in two directions to rewrite
Those errors…amputate, re-draft…a better score
For offspring to conduct.

(The unsuccessful died:
The bowl was shaped as much by failure as success.
The story of a statue's provenance
Lies in the chippings scattered at its feet. So too
Time's arrow pointed backwards to amend,
'Here be dragons, not this way…
I tried, succumbed, re-draft the map,
May you succeed: Keep climbing up')

This two way street of spiral loops
Has 'time' and 'now' in partnership.
The cell erases those small flaws
In chains of 'relegated' DNA.
(They call it 'junk' or 'fossil' anyway)
Shelved in periodic crystal files,
(They might prove useful on another day)

Time's double arrow built in parallel [71]
The twinning rosary in every cell

Involution

Telling, like the beadsman on St Agnes' Eve, [72]
The frozen fingers of our future past.
One chain, the harness of descent,
Drops our emergence from our webbéd feet…
Yet implicate it casts its shadow handicap ahead
(Amended by disaster and success)
To draw the future out of fertile dream.
The other chain is in reverse, to lift us up.

Mind is the intersection of has-been, to-come:
Impulse its dictation, then in action stamped.
Memory folds 'inwards' what was once 'without'
To store in layered depth the all-at-once.
The legacy of past is all we note, yet
At every point of now
The future summons and directs our path,
Recording in another coil what's yet to come,
Potential newly rises with new cultured yeast.

(Aroma of bread baking arouses appetite)

Time is an illusion we adopt
From half a mind, and half a molecule.
First causes we debate, divine or chance?
But final cause is never seen.

Let's stretch our legs beside this bubbling stream…
Spilt from the shoulders of those blue-smudged hills
Gathering as it flows a growing urgency
The contours of its converse with the mud
Repeats, repeats, the sea, the sea.

Lamarck believed a blacksmith's son
Would sport strong arms, a fiddler's child
Be born with callus on his thumb.
He wagered only on the physical
Which, (as is its wont), would prove him wrong.

Canto the Second: Memory, Dreams and Separation

Lamarck, the hero of Soviet rule,
Of ten year plans to perfect the State
With hierarchies of perfectly bred
Political minds for physics atomic,
Hefty shoulders for scything wheat.

*

This pippin that was plucked, condemned
A lonely banishment of thought to outer things,
Until the curtain of the night fell full
Upon a wasteland of destroyed belief.
(All Gods and Spirits gone, mosquito fays
Sprayed with the aerosol of rational doubt…)
The path from sunlight of an Eden song
Was through the charred remains of skeletons.

Instead I seek to show
That Deity rode shotgun just behind the eye,
The quiver of an eardrum, the prickle of the skin…
Impulse in each moment to survive…
Captured, double entry, in exacting nuclei.

*

Darwin and his band of men, constrained
By *'a posteriori'* bands, and throwing bones [73]
Define the links comprising H2O,
List soluble salts, its rate of flow…
Reflective properties, the mountain's peerless lens…
Essential contribution that empowers cells…
The structures it supports, its tensile strength,
The perfect levels it assists, but none of this
Knows aught about its soul that brightly bubbles home.

This is an illustration of my point.
The 'outside' of a thing is not the whole.
Half of all time, the past

Involution

Is only half the tale, (however told)
Evolution is the outside, a mere husk:
The kernel is the consciousness that rides
Back to the beginning, through the mind. [74]

Involution is that ride. My tale now set
To chronicle a newly flexed hypothesis. [75]

*

So now you have stripped bare
The seeming vacuum at the heart of change,
Where progress lacks acknowledgement
(An ill-clad lackey never introduced
To company that keep polite, but never to and fro
Talking of Michelangelo…) [76]
Will you forceps lift that spinal thread?
(The asymmetric serpent to the brain) Now
Pray tell the other tale that may unite
The outer and the inner, mattermind.

For Darwin, mind did not appear
Until a monkey first denuded fruit…
Evolution bumbled, banging into banks
Where 'selection' played the snouted crocodile
Buried in the mud.

'Selection' as conceived is but imposed
(A blind axe to the neck prostrate)
Nature's shuffling of its chosen deck—
(The hand is dealt by courtship's hurried thrust…
The wayward egg survives, or rolls and spoils)
Change that and all is changed;
'Selection' is the new experiment that springs
From inner impulse and its consciousness.

The start of time was fixed when concept God
Turned in insensate sleep to ask

Canto the Second: Memory, Dreams and Separation

What might I be? To set in train the light that pricked
In quanta bandaged up in bonds. Millennia passed…
Before atmosphere arose and tears of water spilt.

Charges now baptised 'electrical'
Bonded that initial dance, enchained…
Danced on in base pairs, D and RNA,
Chemically knotted protein lines…
(Remain electrical in Einstein's brain) [77]

The complex now a Rubik cube,
The hidden cogs more than separate parts…
Until the cell enclosed itself [78]
In a double membrane fort
To repel, unerring, all outside assault.

Conscript minions were enslaved
Under head chef, mitochondrion, [79]
To fetch and carry, tend to alchemy;
The garden of the streaming sun turned green
In the kitchens of the chloroplast.

Each turn of the screw secured advance,
Every single note of the score was stored…
The secretary scribe in her pince-nez
Duplicated the record (just to be sure),
Coiled the ladder round her wrist and filed it in the nucleus.

The two way street was now full flow
Controlled by radioed command, [80]
'Go forth and propagate yourself'
'That way is blocked…'
'Timely riposte. I'll redraw the map…'

The symphony was improvised [81]
From piccolo to timpanist.
The geologic raag entombed,
Performance petrified in genes;
Each emerging voice sang twice, solo and accompanist. [82]

Involution

Some progressed, some fell short
To fertilise the forest floor…
The slime mould's inauspicious name [83]
Conceals its ingenuity; to cluster for defence or sex,
Otherwise content to creep, along in single solitude.

A pin-head size, a single spore,
Takes a collective chemical slide,
When deprived forms head and stalk,
Becoming a 'system', transient, just
To liberate a cloud of more.

Dictyostelium, a single cell,
Carries its culture in embryo,
'Seeds' it to grow when food is scarce;
Agriculture is without aid of a plough
(A portable picnic of egg and cress)

So a small amoeboid cell
No 'higher' than a mushroom's breath,
Keeps tally on a 'global' threat.
Already… lacking neuron-bit
Spurs are chemical, all inbuilt. [84]

You are being parochial, talk about space,
Widen the reference…
Ethnocentric is so old hat.
You seem to be forgetting that
Bacteria still refract [85]
In clouds of interstellar gas…
Survive volcanic chimney stacks,
Boil in plasma, and took on
Tyrannosaurus- no more of Rex-
At a meteoric stroke.
If you doubt this cogent fact
Go seek the signed iridium.

Canto the Second: Memory, Dreams and Separation

Even the man who unwound the spiral [86]
Gazed at the stars to comprehend
Why intermediaries were lacking,
Why no Archaeopter-hiss [87]
In the filo pastry strata?
He proposed Panspermia:
(That granaries in outer space
Seeded new genes on this well tilled planet)
Accounting for the quantum leaps,
The unlikely cell…
The finely focused lucid eye…
The congruence of bee and flower…
The happy correspondence twixt
Infections and their remedy. [88]

(Amazonians extract curare.
Sophisticates inhale cocaine.)

Things no wiser than a virus [89]
Infected earth at intervals,
Raining down in meteor showers,
Bacteria, amino acids,
(Disposed towards a left wing bias, [90]
Doctrine followed ever since)
All the oxygen life needed…
The carbon cycle stabilised…
The organic benzene ring on finger…
Were delicately hammered out,
In Prometheus' sparking forge,
By algae older than the earth.

Was Mount Improbable ascended
By something that had man in mind?

Microscopic is the essence,
(Minute eggs hitching a ride)
Packed in stone on space dispatches…
Accounting for the myriad options

Involution

Of the insect multitude.
(More numerous than any others,
Self sufficient, keeping somehow apart;
Rasping cicadas, diligent ants,
Flying food for birds and bats…)
Such creatures were mimicked by computer   [91]
Fed a slim diet of element genes.
(Intended to prove no need for Designer
By a better designer in the hot seat)

Since when was information uniquely the province
Of the man whose salary is paid for doubt?

Don't start me on spiders or octopi, or
Why the King Coconut bleeds real blood.   [92]
Penicillin has its uses…
Morphine can allay our pain…
Nature knows about its stem cells
(Micro engineering's been around.)
A comet could keep life deep frozen
Until a thaw from passing planet
Let fly the sling. Intrepid David
Aimed stone-shells at forehead Earth.

Careful now. You are becoming emphatic,
Exposing your Achilles heel…
I merely suggested a wider perspective
To grant to man the bequest of stars.
I think that you have well established
That small is beautiful, always was:
From a narrow choice of letters
Shakespeare shaped celestial words.

I'll say no more. Pray please continue
Letting fly time's feathered arrows…

Gradients alone can summon cells
To aggregate or to disperse. Unequal light from either side

Canto the Second: Memory, Dreams and Separation

Build black-cab taxes that traverse the back-street alleys   93
Through sunlit square to darkened crypt.
The 'knowledge' mastered, then inbuilt.

Starlings need the company
Of parents who have gone before.
Other fledglings open wings, setting solo South South-East.   94
The night sky field of stars rotates
In sat-nav brains of novice storks.

The first cuckoo squatter that settled
In the nest of a dutiful bird,
(Completing the décor, out gathering moss)
The opportunist thug, outrageous…
Cuckoos now, with two note scissors, cut instead the ribbon of spring.

Audacity sparked counter measure…
Warblers take to throwing eggs,   95
(Before the danger of delay)
Opposing artillery was set by decision
For both parasite and catch-up victim.

Nothing contentious in any of that.
So where was God?
Gone back to sleep?

It *is* contentious; will be disputed,
(Lamarck shipped out with the Soviet ark)   96
A blacksmith's arms give no advantage to his puny little son.
Behaviour leaves no inherited traces
In physique. But what of mind?

Identical twins parted at birth,
Both choose aviation, or marry fat wives…
Meet for the first time in similar ties…
(Habits regarding trivial tastes
Too similar to be ruled by chance
Or obedient to parental choice)

Involution

How both think, what games they play…
Of buoyant or murderous temperament?
Ballet dancers or banker's clerks?
All set in the aspic of DNA.

The genes are mostly 'junk' or fossil, [97]
Doing nothing, riding along…
Since when was nature profligate, wasteful?
Suppose that these quiescent strings
Connect our minds to everything?

String theory is now cutting edge, [98]
All the invisible universes
Coiled up imperceptible
In dimensions vague to us…
(We are stuck in three plus time)

(The bound slaves imprisoned in the rock [99]
Emerging perfect, though still held
By fossil geologic limits…
Immobilised by straining bonds…
Seem somehow pertinent to this)

The intellect is just as bound
A slave to all its history…
Incarceration in the skull,
Poor Yorick, we do not know him well.

Play it again, Sam, just to be certain.
This theme has many variations
I intend we ornament.

Intelligence, however vestigial,
From inorganic reflex bonds,
The cat's cradle of the spider web…
The wasp's provisions for its egg…
Were re-mastered from the simple links,
In patterned sequence, inner storage…

Canto the Second: Memory, Dreams and Separation

(Involution is the inward flow to more and more autonomy.
The impulse to explore, avoid, is captured, fixed, becomes routine)
A blue tit blustered by a bottle, finding cream for all its kin...
The point of balance on a branch, the span of a leap...
The mongoose mastery of the snake...
Risks first taken in extremis
Embedded in folk memory)

*God back to God to give the answer
Begged by the question he first asked?*

Nobody doubts, watching the embryo,
Formed from two cells, (of surplus swept clean) [100]
Recovering structures of inherited passage,
Retracing from simple cell division
The proliferate organs of earlier kin.
(Creation again, being newly printed
Uses letters perfected, already inked)
Or asks why a man-child coiled in utero
Sports transient gills and amphibious toes,
Resembling a mulberry, later a lizard,
From all the stations of Calvary.

Yet the stages of mind's slow accretions
Are thought lost and never stored.
The central error of our viewpoint
Lies in our inflated pride
In having much too large a head.
Triumphant cortex, sub-divided,
Hemispheres linked to opposite hands...
Yet observe the snuffling neonate, turn, [101]
Crawl four legged, and then stand.

Actions, incremental, trace
The experimental journey.
From the first cell finding another
Through a tail and gradient.
Recombining information,

Involution

From two halves to make the whole.
Child traverses in its conduct
Traditions of its formative lives.
Reflex, taxes, trial and error,
Interchanges, push-and-pull…
Recitation once completed,
The liberated human adult
Constructs new talents from within.

Mind from the sampler is new copied:
From the infant's burbling laughter
(Poured as joyous as flowing water)
Through the studious joined-up grammar
Repeated to dull metronome …
Sung chorus without variation…
Safe in knowing the order of batting,
The queue for cold showers…learning
How to go slimly,
How to get on.

Only mind the costume cupboard
From which to trick out other parts…
Some catch fire (sans invitation) in lurid or in lucid night;
Banked by ashes cold, invisible,
Smothered by pedantic sight.

We were there at the beginning,
(Mystics still proclaim the light)
We were there through all the terrors,
(Monkeys even fear dead snakes)
Chimpanzees climb up trees to foster
Advantage to an injured bird.
Restore a fish, deprived of water,
Cats predict an imminent death.
Elephants circle birthing mothers
With a protective cathedral of careful legs.

Canto the Second: Memory, Dreams and Separation

Who are we to claim compassion?
Dolphins rescue drowning men.

Or comprehend the world about us?
We know its character, not its nature
It is not outside ourself.

*

Let us pause at this plateau
To review the slow ascent.
You have overstretched my muscles,
The saddle chafe, the thrice stubbed toe…
We'll digest the devil detail
Better in a wholemeal loaf.

The lines I draw are truly simple:
Action played the messenger
Between the impetuous single maverick
To the species then transferred.
(Darwin and his incense bearers
Full frocked in academic lace
Outline the mass without an actor,
Priest or Presence at the altar…
Instead a hangman alive to error
Decapitating supplicants.)

Serendipity is too slender
For the weight it has to bear.
Bring on Falstaff, draining a flagon
Or a Hotspur loosing his cool;
Juliet's balcony, a new use for a downpipe,
Malvolio's vanity, several roles for a Fool.
Through actions and actors, mind chose options,
This or that? Either-or?
Cloaked escape rides with terror,
Learning and memory spliced on tape.

Involution

The strata of the mind, like leaves in the fall
Darken, through seasons, the innocent light.
It was the actions that encountered
New solutions, moved apace,
Mastering the world about it,
Storing encounter, increasing discern
On that digital computer
Coiled (needlessly) in every cell. [102]
Each decision chose survival…
(Those that failed, forever gone)

Discrimination was the essence;
Always one thing or another but
Never ever, never 'both'.
'Both' in action is paralysis
(Impulse frozen, certain death)
Every bead of stored encounter
Was knotted divided, 'me' from 'it'
So the selfish mind was severed
From the outer world about…
Australopithecus puzzled
Rolled more 'inwards' from 'without'.

Came the Hellenes, golden polymaths,
(Athens still passed Parnassus the salt)
Round the Agora's evening table [103]
They met to talk of celestial movements,
Democritus' atom, Heraclitus' flux.
All was equally food for conjecture,
All was coherent; all was 'without.'
Mind's mathematics linked it perfectly;
Pythagoras, then Euclid extracted the pith…
Things were either side of equal
Separated by a gate…

(Laws so regal they still dictate.
In the provinces, those barons muster
To underpin the Relative State)

Canto the Second: Memory, Dreams and Separation

Praxiteles, Phidias carving Athena   104
Kneeling in draperies, Apollo nude…
Came rather closer…
In single marble
What each carved was man as God.

Evolution is not only accessible
From bones, or hammers on flint.
In the world of dream and reflection
We visit shrines and gloomy burrow…
See ingenious constructions,
Colours all of dragonflies.
Shake in terror, wake from reptiles,
Weep for what we do not know.
Time is instant, flight is glorious,
All familiar, juxtaposed…
The intellect has stopped its sifting…
Restored the vivid primal light.

(The significant stirs the dream's ingredients
Folding in the daily share,
Linking up events new posted
With long mulched times ahead, before…
Only this, on waking, certain,
Feelings govern. *We* were there)

Our viscous dense aquatic struggle,
Our canopies of leaping leaves,
Our trembling limp through icy pastures,
Our comfort from the woolly fold;
We know from inside we have been there
We empty out the motley bag…
Sleep's laundry maid at the midnight mangle
Pegs out our past, turned inside 'out.'
Men deprived of dream's connection
To the womb that nurtures past…
In the linen press of fabrics
Takes our 'present', interleaves it

Involution

Between the sachet scents of 'then',
Folds new samplers with tight corners,
Smoothes the sheet of future thought…
Men deprived of dreams go mad.

Imagination paints new pictures
On canvas stretched by unheard owl.
Sweating out the fresh connections
From old holes in aching teeth.
Hence we recognise the genius
Celebrating part of us,
Who places his antennae fingers
On the throbbing pulse of truth.

Brain is not the mind's transmitter
Beaming from the summit hill.
More a Bakelite receiver [105]
With valves that warm for 'News-at-Ten'.
It may insulate its cables [106]
In a fat impervious sheath,
Inching out to spider dendrites,
Arc the synapse thoughts that break
Through the sand banks of reception
When they breast the sloping beach.

Brain cells, sympathetic, resonate [107]
Wave consciousness is all about:
DNA is omnipresent
To read vibration of itself.
We receive what we are tuned to,
Science turns a narrow band
Between a tentative hypothesis
Just as far as rational doubt.
Others cross a wider spectrum
From wonder, through worship, and back to belief.

*

Canto the Second: Memory, Dreams and Separation

Messiaen heard birdsong in colours, [108]
Wrote in cerulean or burnt umber keys,
Sought a rose madder hybrid to grind them,
A pestle singing between keyboard and strings.

Van Gogh (at noon) spun a Prussian blue sun,
Melted corn yellow from the straw stubble field,
Drenched in honey a chair of modest demeanour…
Through its commonplace stature, he painted a room.

The egotist Spaniard saw features as fluid, [109]
All eyes (always on him), swivelled and flat,
Three dimensions were one, (and that dislocated)
Guernica, doves, wives a toilette…

Nothing *looks* right; yet everything *is*.

The modest paced day hardly pauses to notice
When the thought of a walk has the dog at the door…
Or the delicate fingers of the water diviner
Tracing bright mercury under the earth…
A farmer observing birds in the thermals
Reads in their chimneys the falling of rain.

These are sips from the glass slipper chalice
In which the immortal lovers drown.

Such wave bands are never or seldom visited
By electrodes fixed to a very still head
That measure only the speed of a nerve path
Or try to recover failing sight.

Slicing, dissecting, describing, dividing,
Mastery keeps its distance from blood.
Thomas invited to finger the spear wound
Established for certain, Christ crucified

(The doubt only served to wound Him again)

Involution

True focus needs a generous platter,
Like those galaxy probes trained on the sky…
Searching outside ourselves for the missing
Connections abundant with shining scaled fish.

Locating the infinite
Gulf between
Two single digits, naught and one. [110]

For the present that space is packed with particles,
All the flavours of shredded quark:
String theories suggest something is listening
Awaiting an opening chance to speak.

My cue, I think, to interrupt
This narrative turning things on their head…
Let's pause besides these sheltering willows,
Spread a rug, have a sandwich, lie out in the sun.
Are we done with didactic? Are you almost through?
Before I bring down the curtain
Is there anything else you wanted to add?

*

The enclosure of mind dissolves when another
Melts the membrane between, the cell of alone.
In abandoning their separate prior existence
Lovers discover the infinite Self.

Lovers of art, or mist on the mountains,
Even cracking the kernels of thought will suffice…
The self forgetful instincts remember
What they have forgot and we know we have known.

Canto the Second: Memory, Dreams and Separation

Eureka certainty proves unmoveable,
Holds the maverick's brain in a vice,
Calls for a strait jacket, and binds the victim
In the prejudice rock, till the world catches up.

Would you proclaim love as the cosmic pneuma?
Dispersed in which consciousness breathes?

Love is the gate through which men enter
A changed understanding of limitless life.

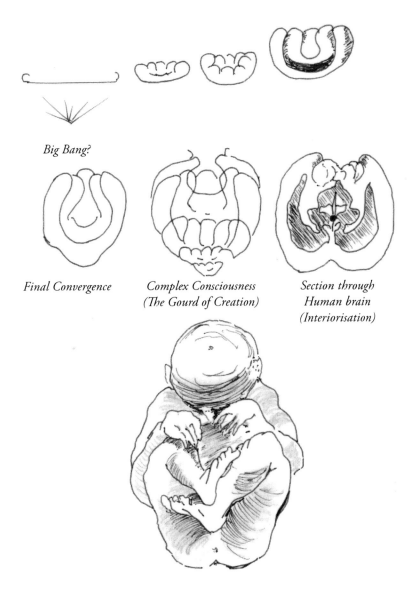

FIG 2 EVOLUTION BY INVOLUTION.
(Prior to Man: Ontcgeny Recapitulates Phylogeny)

Canto the Third: Earliest Days, Society and Language

Do we now ascend a four-in-hand?
A royal barouche, a phaeton
Cantering to man, that miracle?
By your account the summum bonam
Of all that has preceded him.
Conceived immaculate, though not announced
(By angel in silk and bearing lilies)
But unending struggle, spilling gore
Matted in hair, upon all fours…
Yet to stand silhouette high on a hill
Surveying what the devil will proffer—
Dominion over everything.

May I suggest a looser weave?
The garment of all thought still rough
To drape the figure of a man yet
Aspiring to imitate a god.
You are a bridge; upon one bank
The concrete buttress, steely eyed
Must cast a cantilevered arch
To reach the shore of uncertain mud.

Involution

Your urgency to span the stream
Turns rusty from residual doubt
That anyone will follow if
The joints show insufficient mesh.
This Ganges has been often breasted
By those that cast themselves adrift.
Stepping stones did once suffice
To beckon the adventurous.

Come hesitate and test the weight,
Inch forward… else you will blow out
The lamps that flicker on each stair,
Until the sweep of curvature
Lands graceful in the lap of God.
(To unrequited love resigned)
The cosmos is itself fatigued
By cadence frayed, and unresolved.

Where was I? Ah yes, bienvenue!

Neanderthal, an hairy man,
Marvel at his pincer thumbs that peel and probe and nurse his toes,
Cold comforting his upper arms in waiting for elusive sleep.

Very well, I'll take it slow, at a more persuasive pace
To paint the colours, contours, light
In which his puzzled brain explodes.

Freed by cells, maids deputised…
The factory full-steam; the minion molecules a-chatter
Through the corridors of power.

Brain has retired from the fray (marshalling assembly lines),
His underlings, subservient, keep production close to peak.
His part-time hours now released, spent marvelling at everything.

Canto the Third: Earliest Days, Society and Language

Oh brave new world! Ascendant vision
Fronts the puzzled bony brow. Olfaction now plays second fiddle [111]
To the tongue's new epicure.

Yet those hands forever busy, pulling off the beetle's wings…
The pulsating frog on the water's edge, obeying with lethargic leaps…
Termite militia, fully armed, attacking his invasive stick.

The birds escape, wheel round as one
As do the synchronous shoals of fish…
Daily he remains perplexed by the sinking of the sun.

Finds a night spent in a tree, dozing through a barking sleep
Restores at dawn its pallid heat. The future enters like a doubt
To whisper from a distant reef: his mortality.

About him is a patterned strife that keeps at bay the sure demise
By busyness and sudden flight. The bees bed down in a hanging hive,
The lion yawns and cuffs its young.

He is tormented not by fleas, but a new exclusive itch.
Instinctive, others know their place.
Why has that eluded him?

The future is a cracking drought (carcasses alive with flies)
Trees that supplicate bare boned,
Rivers that evaporate.

I'll stop you there, you anthropomorph [112]
(Rückblick ist jetzt streng verboten…)
Time, (they tell us) moves one way
Do not ascribe to the naked ape
Intentions to his idle hands
We know he was only after sex…
A grooming or something to nibble upon…
His emergence was unplanned, fortuitous
In tropics benign, and super-abundant
With incident time to investigate.

Involution

'Time time', you used the word twice
As though it was a cause in itself…
Have you seen time? Except on a clock
Grinning that idiot vacuous smirk…
Time is a construct, as is 'value', or 'charm'
Knots concurrent in a matrix of thought
Time is simply the log of the past.

Ask an old woman racked by pain
Or a child absorbed on a sandy beach
How long is a day? How much of a year?
How deep is oblivion? How distant escape?

The ultimate clock is radio-active, [113]
The amount of matter left over, after
The constant decay of less and less…
Zeno had words to slice paradox [114]
(The closer you get to the speed of light,
The slower moves time; Einstein agreed. [115]
Two thinkers at opposite ends of the twine
Knot the circle of insight…We'll return to that)

I hope I have shown it is 'Action…Now'
The integral impulse sprung from the past
That takes the spur to the reins led on

By a future diverse, not yet on the stage…
Yet no-one doubts its coming certain
Or that today is where it begins.

What's done today perhaps narrows its field [116]
Or splits the roads running abreast…
There are scripts read backwards, or up and down…

Languages we have yet to invent.
The choice is not between matter and spirit
But between man and what he might choose to become.

Canto the Third: Earliest Days, Society and Language

Stuck in the quicksand of what you call time,
Pre-determined by five percent of a plan
Man has yet to unroll the implicit design. [117]

Language is equally frog-marched by time,
The drums behind and whipping the pace
In the goose-step of grammar and sequences.

Time has taken us only one way. Where we stand…
At our feet oblivion beckons. There isn't enough matter left. [118]
Imagination must fill the void with Mverse invisible, parallel worlds.

(All the roads we might have taken instead. Voided by conniving the past…
A balloonist, sitting astride a horse, concluded when he got no 'lift'
The 'infernal contraption' was at fault) [119]

Yet the river in silence pulls on and on…
New turbulent streams drowning the past.
Knowing the fathomless sea awaits
Where all water is water and all is One.

*

That comes later. Stay awhile…
Trace the growth of society, the city, the state
To the brink of emergent philosophy.
Sapiens has done some thinking
But only with a half of himself.
You have laid down the process, the division from 'it'
Soon you will be free to pursue
The slow return from 'it' to 'the rest'.

It occurs to me, reluctant poet…
Iambic pentameter takes its rhythm
From the beat of the heart, lame and strong;
Pump and retraction, like the lung…
Rhythm is in the fabric of flesh,

Involution

The pulse its seconds, a phalanx its inch,
The yard its arm (from fingertip to obstinate chin)
All measurements and time itself
Are man's assembly of himself.

Your theory has delivered up
Incipient man, the sum of creation
By actions that mastered, then assigned
Independence first to cells, then
Organs, nerves, the growing brain:
Brain then mastered strategies,
Learning and storing in every cell
The fall of memory, earlier lives.
Discovery now must take a walk
Through the leaves of the forgotten self.

So keep your rabbit under the hat…
White glove suspense not yet complete…
Wordsmiths melt the rigid word
To shape the soaring dome above,
Clerestories and lanterns open to light…
Light provides the stable arch
Scaffold words are transient…

En-Theos must delay awhile…
Counter the gravity of dominant thought
That believes creation is 'without'
Build the structure up; then knock it down…

*

'Semper aliquid novi Africam adferre' [120]

Eden, first cultured in the Middle East,
With the requisite serpent, dates and figs,
Pomegranates, grapes, ur-citrus possibly… [121]
Apples are a problem, if taken literally…

Canto the Third: Earliest Days, Society and Language

Has been latterly moved to the Olduvai Gorge [122]
Or Taung, Swartkrans, the caves of Sterkfontein,
By men in solar toupees, sandals, and sweat;
Man began in Africa, or so say the bones.

A skeleton of 'Lucy' who stood earlier than most [123]
New incubates the myth in her pelvic aperture.
'Turkana Boy' is elegant, complete, but lacking feet… [124]
Homo now 'ergaster' not only walked: he worked.

Man's pathways have been plotted
By artefacts discarded (the debris of his brain)
In gorges where earth's lips are cracked
To spew out hammers, flints and flakes.

Tools mark man's start: the flagged intent
To influence the future, and assert control,
But what is influencing who? The parchment is unrolled…
The map of the past will be checked against the world.

(Tools have been mastered by Caledonian crows, [125]
Sea snails constructing cases and bright eyed Bonobos,
Man is not, in this, unique…
For the moment let that pass)

But first things first: survival is assured
By scavenge, prediction, hunting, planting seed…
Khiam points for javelins, bone spades in Hemudu, [126]
Adzes in Samoa, shuttles, nets, canoes…

Obsidian in Jordan, basalt in Fiji,
Early man's migrations were in pursuit of tools
Along rivers where he settled, the Great Rift in Africa,
Euphrates in Iraq, the Tigris, Yangzi, Nile…

Earth's fluid flowing arteries would persuade
The solo hunter-gatherer to crouch and bide awhile…

69

Involution

Water led to language and his social skills,
Irrigation, fishing nets, boats and stitching sail.

Man's mastery of water or its mastery of him?

We are mostly made of water; small surprise
That man congregates near rivers, at home with himself [127]
Near streams the villages, 'tells' raised on skulls,
Trade writing was pictogram then cuneiform, on tablets and seals.

From Uruk in Iraq, Halaf and Babylon
The deserts nurtured man near rivers, more tranquil to think:
Sumerians, Hittites, the upper reaches of the Nile
Pushed enlightenment's craft towards Athens et al.

Calendars and water clocks, agriculture, plough:
Rivers and rain were the arteries of trade,
Language, mathematics, ceramics and ores,
Mining, extraction, sailing and stars.

Artefacts and process, smelting, forging bronze,
Are the skeletons of thinking, man's journey through his mind;
The signposts are sequential, from the whole, through the parts…
Back to both ends of the whole, internet and quarks.

For the tabula rasa of 'ergaster's intellect,
The world was an oyster, Pandora's bivalve box.
From the microcosm of himself (Leonardo's circles and squares) [128]
Man compassed the universe to the edge of abyss.

(The pulse of sixty seconds a minute
Summoned the Sumerian, sexigesimal:
He quartered the day of twenty-four hours,
The circle into three-sixty degrees,
Three ten day weeks gave thirty days,
Two Equatorial seasons made twelve months…
The geometry of Euclid
The analemma of the sun [129]

Canto the Third: Earliest Days, Society and Language

Was traced by a finger from the throbbing arm...
The temple created from the beat of his blood)

Divorced from his unconscious Self (one with all)
He transfers his understanding to his 'science of the world'
His conscious mind, fragmented, builds a model of himself
Calling it the Universe, implacable, removed...

The chronology of thought started with the whole
Confusion of the seasons, movement in the planets
The rhythms of the rivers, the vicissitudes of drought...
How could the fragile creature see those patterns in himself?

By breaking the mirror to find levers and gears
Man became omnipotent over something 'else';
Mastery ladled consciousness (the enduring cosmic broth)
Into the slim vessel of his collective intellect.

Evolution, before man, by degrees rolled 'within'
Creatures' encounters with the challenges of life.
Man, mastering language, would reverse the trend...
To involute collectively, rolling memory 'out'. [130]

(Envisage an hourglass full below:
Stockpile of sand, time's memory store.
A single species sitting in the narrow neck
Will upend past into future, exploring himself.

The grains that would start the flow into mind
Would be most recently entered, his emergence from all...
The vacuum confusion, perplexed by the whole
Impervious caprice into which he was born.

Then making slow meaning from seasons and floods,
Celestial travel, the path of the sun...
Vision, that improving extension of brain,
Now begins di-vision of man from the rest)

Involution

Making tools are the actions which slowly transfer
His intuitive grasp, to confirm, certify
That the mathematical rules which created his mind
Defined relationships everywhere else.

Action, once impulsive, continues apace
In man with intentions to master, control.
Alone of all species nurtured through time
His conquest of nature is war in himself.

Opposition built the ladder, all the way up
Encoded contra mundi in mirror based pairs;
There is a kundalini stripe from base to the crown, [131]
A daisy chain of hydrogen, known to explode.

(The cloud above Hiroshima, that mushroom-shaped tent,
Is one application of mind's perilous bent.
The lotus, many petalled, the gold haloes of the blest
Are hydrogen's conversion to other kinds of sense)

Man has collectively moved back through time
From himself to Himself, through intuitive men.
They knot opposing theories into eloquent embrace
Of larger concepts of time and space.

Leaving Relativity tilting Quantum event [132]
(The black-hole knight lancing virtual foe)
Mounted, each is expert preparing for the joust…
In collision neither wins the garter: both capitulate.

This 'event-horizon' is insoluble; oil and water divide…
(In the world of partial relevance these would be combined…
Montagu and Capulet would stand before a priest
Who would fold away his chasuble to receive a Nobel Prize)

This is now getting serious: we're talking Universe…
Where mind is wrestling with itself by looking in a glass…
Mistaking its own image as that 'separated' world
Where image and reflection are left and right reversed.

Canto the Third: Earliest Days, Society and Language

Time stands still, and then runs backwards…
Anti-matter is invisible, or it's dark and not enough
To play Atlas on bended knee, holding up the globe…
Of all the laws and constructs that hitherto have worked…
There must be explanation, if we knew where else to search…
Higgs' boson might slow the leak (look under the yodel, behind the cuckoo clock)
As did Omega minus…at least for a time… [133]
Sorry, scrub that, time's running out,
(Classical Space kidnapped Sabine Time [134]
Herr Professor's rape of our innocent view…
Appearances depend on where you choose to stand,
The speed at which you travel, who else is looking too…)
Electricity, Magnetism, all have been spliced
We're left with lusty gravitation bending over light.

The problem is solved if matter and mind
Are seen as two sides of a talent or coin. [135]
The conscious mind (head intellect) is quantum mechanical, this or that.
It cannot help oscillate for that's how it's stamped.
It was built by di-vision, 'I' comprehend 'it',
The 'tail' options (dimensions to which we are blind)
Are concealed by the choice to fall certainty up…

Consciousness is relative depending how close
To the light the intellect loves and dissolves.

Artists, mystics, intuitive thinkers,
Submerging naked, alone in the sea,
Surface with pearls of infinite lustre
To illumine escarpments and nourish our dreams.

They seldom speak,
(Words corroded by usage are rusty, corrupt)
Mathematics and language express mind's division
Between what *is* and what is described.

Involution

You're doing it again, exceeding the reach
Of logic's reluctance to risk a leap…
The quantum mind steps left, steps right,
Loathe to anchor-weigh until
The juddering pebbles rattle the keel…

Lay out those bones of tentative thought,
The history of tools, chronologically
Knot fathoms as you feed the line
To trawl the journey through memory's net.
Land the catch, keep on ice, set sail for return…

Leaving the landscape to open out…

So I shall, but tools alone
Tell nothing without associate thought,
Anymore than patella will recast a knee
Without femur or tibia needing a hinge,
With a gait and a pelvis to balance the weight
Of a cranium enlarged and a flexible spine…
Each conjecture in the unified field
Articulate only in context of all.

Let me first wrap up the embryo man
Coiled within time's uterus throat…
Yet to deliver himself by degrees
From his history; his sinuous weaving strides
Against combat, hunger, seasons, ravines.
Sudden cast up on plains of grass…
The galloping herds, the rivers' swift course,
Leisure for reflection, of himself and thought.

Tools are what he finds about, and undertakes to make,
Planning for a future, a hunger not yet felt…
The future now takes the rein, commands his mental act
To anticipate the possible, forestall its consequence.
Sifting small portions to his vacant intellect
Sharing the stored larder identical in each.

Canto the Third: Earliest Days, Society and Language

'Recognition' is the word the intuition fits;
'Cognising' once again what he already knows…
Why else would names resonate, last, or be transferred?

It was 'act' and 'acting' that brought man to this pass;
Act as thoughtless as a dry leaf takes the wind
To spiral upwards in a climb towards light,
Delivered, silent on a peak in Darien… [136]
(No Homer to unearth, his odyssey intone)

He has only himself with which to understand
His place in the Universe now beyond
The wall of his cornea, the roughness of his tongue,
The barrier of leather skin; his pondering sense
Of a world as the enemy he has to subdue.

He knows it all, but severed from it, knows it not
(The substance of his nature, one with all)
Dominion over all dictates his mistake
To model the outside from the structure within
Which, exiled, seems uncaring and to him, inviolate.

Tools dismembered are the bones that trace…
From fortuitous flint, here and now…
The wheel, the plough, the water-clock of flood,
The plumb line measures planet's path across the sky
To reach the aerobics of the abstract mind…

The mental life of man's ascent
Reads the parallel transcript in his cells:
Opposition, recombined, transcending strife;
Two threads had long been woven, 'me against it'
(The thesis ear delivered pierced, to hang antithesis)

Tools weighed the worth of fluttering vision:
Tools the means of man's ascent…
From fear of the unknown, excavating the known
Out from the well of unconscient…

Involution

To the scales of the balancing intellect—
Prison'd by reason, and starving the soul.

Introduction

Fig.3 INVOLUTION AFTER MAN.
(*Building the Collective Intellect - Science*)

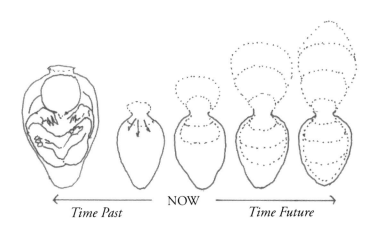

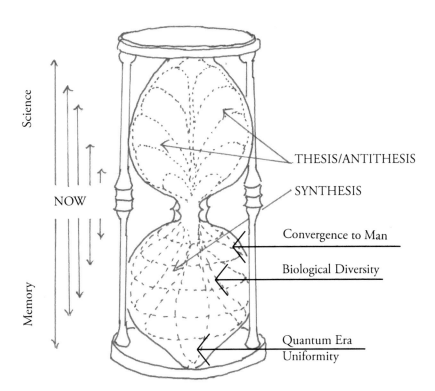

Canto the Fourth: Greece and Some Special People

Hand me the reins, we will out-span
For we have reached the Rubicon. [137]
Release the tired beasts of frantic pace,
Go set a tripod for the Billy-can
(I may use it later for a Delphic perch) [138]

Step out and stretch your spine,
Release the muzzles now all flecked with foam,
Hobble broken knees released to graze,
Unpack provisions for a lengthy pause...
Lo, I have prepared a wide pavilion...

Beneath that silken tasselled tent
We will peruse the history of thought:
I will conduct an auction of the lines
That stamped the border crossings of the mind
And celebrate the gifts of other men's ideas.

(You have read quite widely, I admit
Eclectic is a word... not always kind...
But always for the sap but not the pith,
Digesting little detail, name or date,
The sweet nectar of idea alone was sucked.

You dipped your watered brush in pigments raw
On many palates, disciplines, regimes...
From physics to psychology, then on
Sipping shorts from philosophic tomes,
The formalin of pickled organs, bones...

Involution

Spread the wash across a thousand virgin sheets;
Your colours run, or fade and oft obscure
The clearer portrait of your passage bent
Upon a single search for deepest sense…
The needle in the haystacks of the mind)

I will now use you, merely as a quill,
The spider conscript of didactic fact,
Harness you to stable ink,
To sequence out methodically
The slow progress of the inching snail of thought.

Let's be fair, facts are not your strongest point…
Yet we have reached the hinge of argument.
I have approved your tortuous attempt
In placing mind as coachman and pursuing groom
Upon the lurching progress up the slope.

We now have reached the Rubicon to cross
From arid catalogues, external facts,
Towards the fluid melting pot
Where mind and matter intersect and both
Concur in light's alternatives to glow.

The caravan to take across must first
Be loaded with sufficient packaging of thought;
(The hemp twisted ropes of straining rein
By opposition and refinement soaked
In brine and oil of reinforcing strength)

They have brought us to this deep impasse…
The model of the universal mind
Is cleft upon a gorge, too wide, too deep,
To leap it blind, without the safety sling
Bequeathed by other thinkers we esteem.

Agreed. This is more circus than a calm marquee,
Your trapeze affords advantage that I lack:

Canto the Fourth: Greece and Some Special People

Your noble shelves are Bodleian and stacked   139
Across space-time. I stand below
Amongst the bidders to observe the perfect flip,
The intricate linkages of fertile thought
Across all men, from that high vantage point.
How may I help, if you need help at all?

Go search the cellars and the dusty vault.
You must play scout, the auction's acolyte:
Secure the tool that seals the focus thought,
Hold high the creaky astrolabe, the rusty implement…   140
I'll take the bids and note the surreptitious nod,
I'll give the provenance, you the sequenced shard
To trace a journey from initial Whole, through parts.
The syntheses will ring in names well known
From dusty parchments, tolling bells through time.

Like Christ Church's ever patient Tom   141
That waits for Oxford's scattered carillon
Before it tolls the midnight's final gong?

The cat's cradle of the matrices of thought
Are knotted on a few remembered men…
We can but sweep the mental torch
Across the faces in the building's stone,
Pick out the features that distinguished them
From forebears and head-shaking don;
Celebrate their long shadows cast,
For they were giants over lesser men.

Mind is immortal but unseen:
Unlike the vast and trunk-less legs of stone   142
That should attest to King, but to decay succumb.
The sands of level thought are never lone.
The perfect circles that were thought Empyrean
Remained until poor Kepler stretched at them   143
And found ellipses, in contrast, a sad 'cart of dung'
(Parabolas would, in due course, rescue them

Involution

    From ignominy; revealing greater unity beneath—
    Gravitation pulling out the plum)

    Harmony was always guided to true North;
    Deflections were resolved by deeper thought,
    The compass moved beneath the needle point
    To re-calibrate perfecting nature's God.

    There is a glorious disparity,
    The paradox whereby the caravan
    Was led by travellers, uniquely unafraid
    To find established emperors had no clothes.
    They dragged all cheering, jeering men behind
    To gather up and measure what fell short
    Below the turning spokes; securing paltry spoils
    Of disputation and the fashioning of hats
    From small distinctions, narrow ribbons of reward,
    Proving them mistaken, finding brighter claim…
    (Perfecting with triumphant diadem
    What was, in general, adequate au fond)

    The penetration of the well of memory
    Has been collectively transferred,
    As were initiatives in evolution heretofore…
    The maverick impulse embedded in the mind of all.
    The act was now the act of thought itself
    By men too big for boots' conformity…
    Tools and the checking of the cogent facts
    Proved the inspiration founded and, in essence, just.

    Some tailoring of language; taking in the slack,
    Stitched garments fit for all then to adopt.
    Intrepid thinkers handed on the torch
    To re-ignite receptive minds in later 'cogito' [144]
    Always it was single thinkers who escaped
    The weight of all the reasons to say no.
    This is the salient constancy
    To persuade the doubting now to solo fly,

Canto the Fourth: Greece and Some Special People

In company with famous pilots without legs ¹⁴⁵
(Red Baron in a Fokker, Brothers Wright or wrong)
No canvas tethered to steel cleat
Ever hopped a hedge

*

Enough said. Let me climb…

In the early uplands of creative thought
The world was still an oyster, integrate…
The Gods, the stars and planets, small concerns
Of governance, morals, manners, all enclosed,
All food for abstract thought.

Before the Greeks no individual named
Could claim Seleucid knowledge of the naught, ¹⁴⁶
Or the Egyptians' battle with the Nile
Flooding annually the careful valley ploughed.
Geometry restored entitled calm to angry men.
The Ionians were unafraid to think ¹⁴⁷
How patterns might emerge from thought itself…
Thales travelled widely and observed
That water was the 'essence' of the world.
'Not so' said Anaximenes, 'it is air' ¹⁴⁸
'The pneuma we all breathe unites…'
Hecateus pooh-poohed amorphous myth
And mapped instead the solid certain earth.

Heraclitus wrapped it all in 'flux' ¹⁴⁹
Nothing was either born or dies, but moves…
(Lavoisier's 'conservation' was already there
In transformation; air fed fire: liquid water, ice)
Democritus, preceding Dalton, rolled the dice ¹⁵⁰
Of atoms 'uncut-able', underlying all…
Herodotus set off impatiently ¹⁵¹
To set his seal on Anthropology.

Involution

Flux in ideas, vicissitudes of thought
(From microscopic to entirety)
The opposites combined, and re-dispersed,
The Whole was macro large and micro small.
New syntheses required later men…

Pythagoras of Croton reigned supreme: [152]
He found in 'number' now the magic sun,
The abstract underlying justice, music, reason and the march of time.
Ten fingers, one and zero, ten the pentacle,
The nine planets thirsted for a 'counter earth'
(Like anti-matter, virtual boson now
So the Croton's mathematics then
Invented what was perfect and unfound)

Shall I take round the monochord still strung
With single cat-gut taut? And show
How half the length, when stopped and plucked,
Sings out an octave higher than the whole?
Halving it again will give the fourth,
Smaller fractions follow other fingers' width.
Number was aesthetic, pleased the ear—
Its music was the concert mind of God.

(Harmonia, the music of the spheres,
The structure of the pleasure of the heavens,
On court balcony, in rustic jig, the key relates…
Proportion gilds the tapestries of sound
That moves the heart, or tricked by cadence held
Resolution, unexpected, laughs out loud.
Until the equal lengths of Schoenberg's twelve tone scale [153]
Smoked music for dry intellect as well)

Shall I resume? Not many tools
Could measure the conjectures of those minds…
Perfection was the only guiding light
For any bowler's open wicket turf…
Philolaus early knew that earth did circle sun [154]

Canto the Fourth: Greece and Some Special People

Although Copernicus, rather bashful, published first

(In plain brown paper to selected friends,
Unwilling to expose untimely thought...
It took three hundred years to find a press)

I have broken your chronology...
In Greece no limits were imposed;
The narrow boats of mind were not yet oared.
The Aegean sponge with seaweed floated free
Dispersing Greeks to fairer Italy,
Egypt, the Turkish coast and Sicily.
Pre-Socratic thinkers, as they now are known,
Were each possessed of virgin, untilled loam
To seed with new conjectures, grafted shoots...
Philosophers expressed ideas as poets,
Incantations that were known to heal...
Logos meant just what it claimed to mean,
The spoken word as precious as a stone
Carved upon the air with the first speaker's name.

With nubile daughters of the sun, Parmenides
Was admitted to the unnamed Goddess court,
Transcribing most precisely what she said
About the only road (less travelled by the lost)
Towards eternal life: 'to die before your death'.
Plato, jealous, later stamped her book...
His 'ex libris' smudging what she meant.
Thus the first messiah who offered life eternal
Became another Father of a Church
Of Athens and its commandeering Schools...
Remembered now for 'rationality'
When what he sold was guided ecstasy. [155]

Empedocles ventured into matter:
His 'roots' of earth, air, fire, water...
Each soft 'element' with willing inclination
Attended Aphrodite whipping up corpuscles...

Involution

Her foamy surf washing all ashore…
(The beguiling Beauty fashioned from deception
The world revealed, so easy to seduce) though
Love promised what it never could deliver,
(A creation severed from its Source, its Unity)
A growing thirst, denied, would shape the longing…
Carve cheeks to gleaming gaunt and sinking sallow…
Press-gang pained youth to shiver, growing wasted…
Take leave of body's coat and empty pocket…
And shun (through Strife) all Love's alluring song. [156]

The rigging of the ship of consciousness
(Picked out against the dark and ebbing sea)
By flaring lamps suspended on a cobweb,
Each taper lit by Herculean men… [157]
These disciples of Aegean Galilee
Sat at First Light unknown to one another [158]
To breakfast on the business of creation,
Share out a Xifias of being and non- being: [159]
Opposition braised in basted somersaulting…
Prod logic's limits, slice through fallacy…
Sweet cordial they sipped from immortality.

(Mark well this fragile port; we'll slip our moorings,
Our laughing Temeraire carried by the tide. [160]
When we have sailed into the thinking future
'Tis here we will tie up in time to come…)
Observe the entertainments here displayed,
The spoken word, the chorus and the song,
The lone Hippocrates, wandering the hillsides [161]
For herbs that smell of sunlight, and of lime.

The theatre where Aristophanes explored [162]
The human heart, clouds, laughing frogs or birds…
History was in poetry intoned,
Science was the science of the mind,
All compassed equal, merit and acclaim.

Canto the Fourth: Greece and Some Special People

The Golden Age, so named, all qualities of man,
His passions, music, sculpture, poetry
Were equally the attributes of Gods;
Familiars who sometime graced the table, but
Burnt enigmatic boats in Delphic smoke.

So let us now to Athens all repair
Where riches now await and 'schools' take root.
The stranglehold of Aristotle, the new noose
(A leather bridle that will come to gag and choke)
Yoked logic, system, hierarchy...
Bound kingdoms and anatomy...
De Anima, Physics, and placental lamprey
(Not confirmed until the nineteenth century)
Hawk-sharp in his eagerness of eye
Gobbling, grasping everything that moved...

Stay. Not yet.
Let's peg out all Platonic virtues first.

Before you hasten to the great Divide
Let us observe the cracks that first begun
Within a single name: Hippocrates.
He of Cos gave scrutiny (now look again) all sway,
Conjecture was humanely drowned at birth.

He of Chios on the other hand [163]
Came close to squaring circles in his head...
Circles were as the squares of their diameter...
Lunes related curved planes to rectangular... yet poor sod,
His royalties were commandeered by Euclid.

I am surprised you did remember that...

It was the mere co-incidence of name;
A single 'I' in place was all it took
To separate the parting of the ways.
It was thereafter ever thus, and since

Involution

This is a central lynchpin of this work
I could not let it pass without remark.

*

Do I regain the floor?
Although the concert of the progress of the mind
Was played by names that bell-like toll through time…
The music navigates its onward flow
From rock to eddy, stagnant pools and on…
The headwaters of Hellenic gushing spout
Revitalised all allied later thought.
The journey that impelled, and then confined,
Implicit in the philosophic start
Where science, virtue, conduct, coalesce
In thinkers who believed themselves the thought.

The full and balanced meal will now be carved
By narrow mastication of the part.

(Be not therefore easily diverted
By any single name or small concern:
Stepping stones are merely there to serve
The progress of the exiled man's return)

Now a prayerful attitude is sought
In crocodile of tonsured heads, well shaved…
As we approach Academy illustrious [164]
(That spawned ascetic imitative shoots
In Benedictine vow and Oxbridge dormitory) [165]
For ambulant debate, in shady cloister
The 'ars vivendi' was to study for itself.

Across high noon's flooded marble terrace
Scarabs reign, incessant crickets sing…
Awaiting evening's passagiata, when
Peripatetics take the kinder sun, [166]
Pouring ouzo, raising heavy tumbler…

Canto the Fourth: Greece and Some Special People

With stony sightless eyes Pallas Athene   [167]
Holds up the torch for architects of thought.

The keeper of the keys, old Socrates [168]
(If he was ever young it was forgot)
Ironic in sardonic disputation
To prove the Delphic Oracle mistaken.
He failed; fell in a swoon at Potidaea,
Came round much changed and evangelical,
Exhorting all he after met to heed the Soul:
('Know Thyself' his central admonition)
Life's solemn purpose was its liberation
From the prison of the mortal coil.

Immortality could be gained by contemplation,
Moderation, justice, discipline…
As you pass his stoop take a libation,
Dress in sack-cloth; mark your brow with ash.
Sipping bitter hemlock, seated, tranquil,
Serene and unresisting, welcomed death…
The first martyr for an independent mind.

After Socrates we take a road to hell.

The Great Divide now becomes colossal
(Shadows cast by two Goliath men)
So let's unfold a table in Agora, [169]
Drink wines of lustrous reputation…
View smaller vineyards planted round about…
The Academy exhaling softly
Shadowed by the thrusting new Lyceum
Where Aristotle prunes and paces…
Pedant pupils hastening behind…
To gather up the smallest clipping doubt.
Ever after, both estranged but marching
Two clear sword lengths apart: Since then, [170]
All men, it's said,
Are Stagirite or Platonist at heart. [171]

Involution

(Which am I? Or need I ask?

You are, as yet, unreconstructed...
Even Plato would your suit disown...
Plotinus might accept you later [172]
But we haven't got to Rome)

The Doctrine of Ideas is central
To Plato's concept of creation,
Ideas were absolute and all.
The soul which (unmoving) moved sensation
Composed of perfect concept, 'form'.
The Cave (in which we are asleep to warm our cockles)
We misread in rough unpolished mirror
Illusion flickering shadows on the wall
Cast by our heads; our backs are to the flame.
Only soul restores our final port
When manifested matter is burnt out. [173]

The closest match was mathematics
To ideas and their unchanging form.
All mathematicians are therefore Platonic
(Matchless Euclid still the paragon) [174]
Logic and deducing limits,
Dialectic and induction,
Mathematical gymnastics
Keeping oiled Platonic marbles...
For light relief he wrote his 'Laws'
Timaeus kept vague the esoteric
In warm straw to ripen slowly...
Until Aquinas plucked his perfect God. [175]

(Aristotle was more the ticket when Aquinas
Offered remedy for far-from-perfect men)

Plato has fed mind's adventure
Into thought and thought alone...
Consciousness the trumpet herald

Canto the Fourth: Greece and Some Special People

For solemn seating at high board:
Virtue, Morals, Jurisprudence,
The noble Arts of Man and State,
(In contempt all acquisition
Pomp or power, influence…)
Giving cursory attention
To nature or ephemera…
For those we cross to the Lyceum
Where Aristotle makes amends.

If 'by works shall they be known'
Aristotle is supremo ultimato…
He first taught young Alexander,
Packed him off to conquer the known world.
Then settled down beneath his olive orchard
To master what remained, all thought.
Started humbly on his modest weeds…
Then the botany of all known vegetation,
(Drawing in a comprehensive 'herbal'
That, to this day, would hold its own at Kew)
Progressing on to octopus and squid,
The dogfish and the embryos of bird…
Before breakfast wrote 'On Parts of Animals'
Dissected, named, sketched the varied views
Of the cow's four chambered stomach
And the ligaments of heart.

He saw the ladder of evolving species
Becoming one from one another
Through the mounting subtleties of soul:
Plants had life but 'vegetative,'
Animals were clearly 'animate'
'Rationality' ennobled man.
Embryos revealed all the connections…
(Darwin, later simply re-examined,
Filled the geologic and conceptual gaps)
Where Aristotle did some damage
Was in celestial mechanics…

Involution

In peerless orbits he stacked spheres,
(A crystal and hermetic citadel,
Where the hierarchic inter-locked and governed,
Laws too remote to reconcile on earth)

His gulf between mundane, celestial,
Stifled speculation for two thousand years…
Despite his stated modest admonition
That science should adopt new clothes
When the old proved threadbare and outgrown.
Instead the void was ripe for planting
By astrologers and proselytizing Church
That bridged man's crippled understanding
By claiming (because they could decipher Greek)
They alone held divine dispensation
To mediate across the unknown realms…
If Aristotle's giant stature suited,
It was neither his intention nor his fault.

This bargain basement tour of Athens
Was cursory to say the least…
I will be ashamed to own it…
All those amputated statues staring
(Without noses or their trophy toes)
Few painted by a passing glance…

We have all of history to get through…
This was to peg the great division
(Which this tract aspires to bind)
If you'd rather curl up with the Phaedo [176]
And sip ambrosia instead of watered mead…
You'll find it all more Greek than usual…
You can join the cloistered schoolmen [177]
Dusting off the fine distinctions
That won't add a hair or soupçon
To persuasion's hill of beans.

Canto the Fourth: Greece and Some Special People

I am satisfied to signal
Why philosophy and science sundered
From all embracing integration
Of macrocosm and the mind...
Two charioteers of equal stature
Drove coach and horses at an angle;
One took the path of observation
(Science now will follow him)
Of minutiae, change, decay...
The other stayed with contemplation
Of fundamental mastery; self in service,
Morals, conduct, harmony
To realign creation's purpose
Through the sanctity of life.

Mathematics staying silent
(Unintelligible to most)
Will accompany both journeys
To the self same destination
Where mind and matter stay divided
By languages and intellect.

The lash that whipped both sets of horses
Were individuals, sole and soul.
All Platonist ab initio,
Contemplative, lost in thought...
Surfacing from inspiration
Becoming later Stagirite,
When words and tools and observation
Were needed just to validate
The connections, laws prevailing
In mind's dark sea and in the light.
Science needs corroboration;
Its geniuses all have to wait
For tools (that may be rare eclipses)
Mystics will reveal through works.

Involution

Have we now forgot our manners?
Leaping to their consequence…
Should we not return to curtsey?
Wash the feet and kiss the hands?

They are all far too busy
Sowing petty argument.
From the champagne of Aegean
We should take a swift depart.
Division now will quickly clamour
To debase the calm debate
Which asked too much of man's obedience,
Denial frugal, virtue stout.

Stop your ears: the strident Stoics [178]
(Pantheistic to a man)
Know-alls shouting in the corridors
'Ether' 'Pneuma' what's the difference?
Sell the soul to Tarot cards…
Necromancer and astrologer
Taking coin for sooth saying.
(Gods were now capricious Fates)

Leave the Romans to rename them
Vulcan, Neptune, Mercury.
Rome would swallow easy options,
Armies on the move walk light,
Stoicism shrugged its shoulder
For most of the Imperium.

Yonder hear the Epicurean, [179]
Not much different, in the main.
Epicurus, he of Samos, deplored methodical pursuit.
He believed Democritus' atom
Was the only game in town.
All was roulette, accidental…
Life as well a bed of roses
What was wrong with women and song?

Canto the Fourth: Greece and Some Special People

Lucretius was of his persuasion, [180]
He would later take an interest
In cataclysmic volcano,
Tempests and all conflagration
Was music of impressive din;
(In Rome with all that method marching,
A man needs some adrenalin.)

(With Chance, Caprice,
Dual drivers at the helm...
Small wonder all were libertine.
We, offered sub atomic boson...
Random error making progress,
Answer in a similar vein...
Flip-top sports-car, fine malt whisky...
Caribbean cruise or pad...
Rolex watches, pure silk knickers...
Drown despair now, just as then)

Peripatetics in small numbers
Still cogitate in cubicles.
After Aristotle, Theophrastus [181]
In his instinct, naming names:
Pericardium, cotyledon,
Botany is to him indebted
For the sex of plants.

Autolycus, Dicaearchus [182]
Make terrestrial advance
From home ground's geometry,
Drawing latitude, rotation
(Aided by astronomy...)
For Pytheas's navigation [183]
Sailing from Provence.

He observed the moon's persuasion
On the timing of the tides,
Circumnavigated Britain,

Involution

> *Took the home run down the Elbe,*
> *Tied fatigue up at Marseilles.*
>
> *All these were, no doubt, helpful,*
> *(Partial in comparison)*
> *Science now will narrow surmise*
> *To the useful and applied.*

I'm persuaded, let's adjourn.

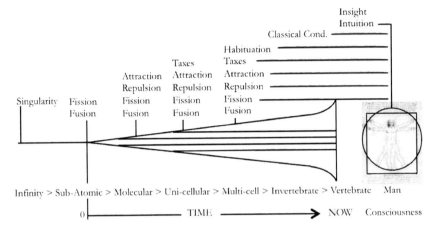

Fig 4a. Evolution of Behaviour- indicative of complex consciousness. Simpler elements are incorporated but all are retained and continue at different organic levels

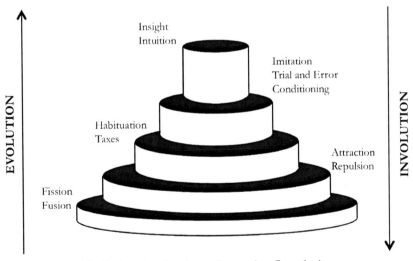

Fig 4b. Acceleration due to Increasing Complexity

Interlude: Archimedes and Alexandria

There is a slim vessel in Piraeus rigged,
A rust red sail is half unfurled
With rope loose-hanging at its waist
To lace the billowing sail above…
We'll seek the harness of the wind,
The kisses of the slapping wave…
A sloop to make full well for Rome.

'Full well' you say; I am unsure
Why Rome commands our reckoning…
Alexandria has a better claim…and
Riper fruit of thought to bite.

I'd hoped we might over-night
Under Ptolemaic rule…
Buy braised goat, fermented milk
Or, better still, a stronger brew
From vendors wearing tasselled fez…
Drift south in a felucca piled
For commerce on the slipping Nile;
See Gizeh with its sightless flanks,
Pyramids pale, austere, remote.

Alex has no ice cold brew…
The torrid throng, the desert hot.
It is an Athenian summer school,
The last polymaths did there repair
As did the Popes to Luberon…
Up promontories to catch the breeze,

Involution

Commanding all arcaded views
Where the corpulent prelates dined
Awaiting troupes of dancing girls…

No: Respite and the briny deep,
The mewing gull, the albatross
Must salt our resolution brave
To furrow-plough the water field
To face the pomp of clashing shield.
We'll harken to the quiet voice
Of poets that in turrets flutter
Like contented turtle doves…

Ovid, Virgil and Catullus,
May weave a lyric melody
They are not our quarry now
What have they to do with us?
To forsake Aegean I am loathe
Until full milked of polymath.

We have a mono-rail ahead
Of oscillation's narrow search;
A barren desert to survey
For first green shoots of Renaissance.

A compromise is hereby proffered
In answer to your deep desire
To visit in the dying days
Remnant Athens drawn to Alex.

We'll travel to Sicily, a fertile isle
With vineyards, goats and rough red wine…
With Heiron we will sojourn well.
It's said he keeps a tethered squire:
In his garden there resides
A Venerable, much more than Bede.
(As Haydn was to Esterházy
A constant stream of world renown

Interlude: Archimedes and Alexandria

Servant in name, but in truth
The jewel in unsullied crown)
So Archimedes serves that suit
For Heiron: he just thinks his thought.

Beneath his much famed walnut tree
We'll take a glass, and reminisce
Of Alex where he journeyed oft
To hammer out untested thought
On anvils of his stature, wit…
The Renaissance begins with him,
In Syracuse Corinthian
We'll catch the last lizard by a tail
Before it drops and disappears.

*

'On the outskirts of the town
Between the buttresses all clothed
In oleander, and dense shade
From carob trees, he will be found. They say
Plaintive goats bleat through his day. White pigeons
Scratch his modest ground. His simple bench
He occupies from dawn to dusk:
Tracing circles in the sand,
Making notes and diagrams
Upon papyrus sheets, procured
When in Egypt he devised
A screw ingeniously raising water
From the river's sluggish bed…
Sluice arteries swift now irrigate the planted banks on either side.

His real pursuit was otherwise:
He sought principles which under lied
Mechanics, (kings found practical)
As illustrations of his thought.
Water more than welcome draught
But medium for displaced weight…

Involution

Fulcra balanced different loads
Depending where he sited them.
He saw with prism clarity
That thought bound all and did contrive;
Inscribing and approaching limits…
A polygon in points was circle…
Nearly, nearly, but not quite…
Squeezed the error, the pip-squeak pi
Corrected all uncertainty…
All wholes were aggregates of parts,
The 'Sand Reckoner', that he called his mind
Could compute the universe.

Shall I lead the way? Who shall I say?
He is recluse. He'll need a reason
To interrupt his mental fast…'

'Tell him we visit from the future
Not tourists through his memories
Sand sifters of a different sort…
With things perhaps we might impart…'

*

'Visitors? I receive no visits now,
Bid them depart. Say I am old
Or, if you will, no longer sensible.
Where are they bound? To Rome?
Tell them to tarry in the town…
Rome is on its way to burn us out.'

Such visits weary, what's in a name?
They'll nibble raisin detail, but tear the wholesome loaf…
All sweetmeats bruised by questioning…
What do I recall of Apollonius? Well named for God.
Will he appear at even, lissom, on a rock?
Bronzed by the setting sun…his chest a-quiver…
Ribs all laced with indrawn breath…

Interlude: Archimedes and Alexandria

A moment later, the ebbing sea enfolding him
Its meniscus thick with molten gold…
But they will ask 'What did he leave? What conquer?
What import did his calculations hold?
Ellipses, parabolas, hyperbolas of talk
Will drown my precious vision of the boy.

'Tell us of Eratosthenes instead,
Was he not Curator in your day?
Did he not put a girdle round the earth?
Was it not his sieve of mind?
That shook out the succession of the primes?
Pierced the sphere on which he stood
With an axis, pole to pole, round which he moved?
Was not the equator what he named
Imagination's scalpel that incised the globe?'

'If you know the substance of his muse
Why ask? Why are you wasting time?'

Eratosthenes, my friend, how desolate
Your absence proves! Remember how
I ran through all those swarming streets
(My slave was burdened like a beast)
So I might leap the Library steps
To your high eerie in the roof…

Yes well…
I might have travelled through the realms,
My continents of troubled thought…
Yes, yes, raised water from the river's bed,
Swung the earth around my head…
Moved the pyramids to tilt
Upon a fulcrum, and submerged
Floating bodies, displaced weight…
Seen planets circumscribe the sun…
Yet none of this answered
The still small voice that whispered

Involution

What is a man? Why has he come
To fathom out all that he is?
But not know what he thirsts to find.

That alone I received
In small inflexions from a hand…
Quizzical amusement in a brow…
A surreptitious fleeting smile…
The shaking of an honest head…
The synchrony of walking pace
In the presence of a friend.

A friend is not reducible
As are ideas mechanical.
What did we choose in one another?
What drew the bond? Distance was in fact enhanced…
The adventures found in solitude
Seasoned parting and renewed
Rich sauces for a new meal shared…

Therein lies answer, yet I full know
It will not prove amenable…
Entire and too delicate
For intellect's analysis…
The heart has its reasons, inchoate,
What do I know? Or know I knew?
I knew Eratosthenes.

*

'Make my excuses, bid them refresh
At Arethusa's gushing spring… Aeschylus
Is between his drafts, he may give what I cannot…
To him is audience meat and drink…
I do best in solitude…

Interlude: Archimedes and Alexandria

*I have nothing else to add:
I am ready now to leave.
Life is empty and fatigued
When there is little left to learn.
My almonds all will grace my grave;
Their kernels hard, small meteors
Will fall like gravel on my tomb…
Like rain they will refresh me there
With season's passing, and the starling
Persistent, clapping shell on stone
Will feed me as I feed the worm.'*

Canto the Fifth: The Dark Ages. Monasteries and Muslims

We called, at your suggestion,
But answer came there none...did you deceive?
I had thought you nigh omniscient, so explain
Why the oyster with the pearls kept stumm?

If you thought me Delphic in prediction,
You have seriously slipped. Time is but a vapour trail...
'Now' is where creation happens; prescience of what may come
Is unreliable, which is why, Madame,
She speaking always with forked tongue...

Look deeper: He received us both,
His silence was his apt reply,
He knew our needs, and our intent
To reveal how mastery drains the cup
But changes no water into wine...
His absence was well judged, perhaps...
We have a desert to traverse
(Twelve hundred years of endless drought)
Why slow it further, digging up names
Of every corpse that ever thought?

Instead let us smoothly undulate...
A snake through dry leaves of the past,
A coil briefly visible, a ripple submerged,
Muscles alternately flex and contract...
Sliding appearing quite serene...
(Ignoring that endlessly flickering tongue)
Progression from tail to searching head

Involution

(It does not haul its length behind
The rear impels the forward thrust)
Just as with knowledge; what exists
Selects the question, next attack.

It is as though spark mind alights
On tinder dried to re-ignite...
Genius seems to find its kin
Across centuries and distant clime.
Athens struck the fountainhead
Water gushing limitless...
Whole man, whole universe as God's
From which the diverse divers plunge
Down tributaries from man out-splayed,
Fragmentation will search out
Partial drifting of the part; all to coalesce
At the single river's root.

In mind the journey is reversed, [184]
Time moves from now towards the start:
From the waist of hour-glass
Knowledge is by tools transferred.
The mental mirror that gets filled
Becomes what's called 'consensus world'

So the past dictates direction, what about
The fertile fields of people and place?
Athens flowered as it breathed,
Alex hooked Euclid on bait,
Apollonius came hot foot,
Others followed up the line...
(Rare scrolls were very good as bribes)
Alex will succumb to Rome, and learning will take startled flight,
Landing as a flock in Persia (scrolls beneath the white djellaba) [185]
Then monasteries on hostile isles...
Where British cells will be austere,
(All those Matins, Lauds and Primes...) [186]
Save my knees: Recall me. Why did I commence

Canto the Fifth: The Dark Ages, Monasteries and Muslims

This traverse of collective mind?

You thought yourself evangelist;
A miner with a richer seam,
A tungsten lamp to navigate
Through darkness error… subjugate
Your listeners to another state…

Have you considered?
This journey's trace eliminates
(By removing signposts as we pass)
False dawns, false doors, dissolving matter…
Your recollections of your search
Are pressed in pages, here perused,
Turn each… it goes up in smoke.
(No man read his way to God)
This is the Eden once bequeathed;
The garden over-turned, dug up…
Our mission is to prune the tree
Of knowledge and its bitter fruit.

With regard to place and crowd…
As hybrids renewed the common pool
(Specialised, pure-bred died out…
A wing was first an upper arm…
Unused it shrank, until vestigial)
Ideas are creation's higher trials
(Professor Reductio calls them 'memes') [187]
One man's idea is tongued by many,
Hybridized, exchanged, refined,
Corners are removed and smoothed,
Sanding follows, then the varnish…
Cabals place it in on a plinth, where
Static theory ossifies.

Come: choose our transport, now appropriate
A monorail? With what swift ease
It will span a thousand years

Involution

Of tundra, scrub, Siberia—
A landscape most monotonous
Devoid of hill or handsome grove.
The oil of Athens will ensure
An impetus quite effortless…
Few level crossings, stations, halts
To alleviate ennui.
Uninterrupted rail was forged
In steel of everlasting thought.
The Stagirite our special smith
(Besides us rolls the parallel
From Plato who will soon diverge…)

I propose a game of chess—
You play all pawns, those marching troops
Like locusts they consume all thought.
I shall play the royal part,
My bishops all will crabwise move,
My knights o'er-leap the trenches, banks,
My castles keep within the fort.
So gnaw with names, your facts approach
My boiling oil, my drowning moat.

A game you call it: well perhaps
The Gods do play with mortal men
In hope we may collaborate
In joint adventure to embrace.
Dividing rails that run beneath
Were granted, so pursuing either
Would, over time, reveal that neither
On its own could satisfy.

(Science of more jointed facts
Proliferates more cracks between.
The fortress, Colditz, where conceit
Dines nightly on thick lentil soup,
Raising never glass or eye [188]
To Eleusis or Prospero.

Canto the Fifth: The Dark Ages, Monastries and Muslims

Blind faith is but a nagging tooth
That yearns for clove of revelation
Or chews on gat of miracles... **189**
Sublimating pinching doubt
By rituals and incantation)

Since to the first we are confined
Is there a bypass around Rome?
I've heard about its catalogues,
It weighty cumulated tomes—
Lucretius on the 'Nature of Things' [190]
Varro on the 'Liberal Arts',
All stoics, epicureans,
Determinist and roping steers;
Anything that bucked or kicked
Brought low and in subservience.
Pliny struck in Pompeii's ash, [191]
Observing at too close a call
'Natural History's' rich contempt
For unyoked curiosity.

Should we not play Philistine?
And pass over on the other side?

Would you ignore the incomparable
Poetic language unsurpassed?
Do you not marvel at that music?
How men on constant march devised
The roots of language so precise
For names and naming of all parts...
Jurisprudence, medicine,
Seeding all romantic tongues...
A grammar in the word inbuilt
From Britain, to Gaul, Iberia, Indus
World submitted (more or less)
To a tongue deserving of temple;
Ciphers more than elegant, speech
Clean and crisp as a vestal virgin?

111

Involution

I know their tools were well advanced
From dioptera to abacus, [192]
Water levels, water clocks,
Cranes and pulleys, pocket dials...
Even Caesar found the time
To re-calibrate the calendar,
Signing, with a flourish name,
July for his and summer's height. [193]

An inspired plumber, Vitruvius [194]
(Are drains and heated baths your thing?)
His Cloaca Maxima flowed (unsweet),
He slept beneath mosquito nets...
(He, of all, knew Roman quagmires
Though Horace mocked his cautious ways)
I've seen pictures of his aqueducts,
The well-sluiced streets, the hospitals,
All politic for conquering...
Surveyors marked the marching miles,
Roads went straight, undoubtedly...
Practisch, doch, I do acknowledge, but
Ordered appearance mostly hides
The lack of more creative chaos.

Rome secured the Greek donation:
In poetry, dramatic art, man was
Simple, central, human,
(Idolatry was not conceived)
God was shown as sometimes naked
Or clothed in peerless draperies...
That part the Romans took, developed,
Architecture looked to Greece;
Phidias, Praxiteles...
Virgil rode in Homer's silk.
Cicero took a Sabbath day's journey [195]
To kneel beside dead Socrates.
Artists wore the Grecian tunic:
Science served Imperium.

Canto the Fifth: The Dark Ages, Monastries and Muslims

The Greek cornucopia of thought
(Heedlessly spilt and almost perfect…)
Too rich a diet for Roman column,
Digestion took five hundred years…
The lyre of Hermes found no ears,
His boundaries alone were breached, [196]
Plundered for edicts by great Aquinas…
(Look sharp, secure him, canonize)
The prime black-hooded 'domini canes' [197]
Then stoked the Inquisition's pyres.

There was a Roman worth a pause:
Plotinus linked the parting rails, [198]
Multilingual, polymath
With a quasi-Buddhist strain…
Stirring Sufism from Persia
In India's spiced and fragrant teas.
He was neo-Platonist, ascetic,
The created world he saw 'en-souled'
(Gaia is its modern name)
All soul returns from whence it came…
Bodies reflect, as a mirror, forms…
Thinking creates, but first must **be.**
His devoted pupil, Porphyry
Swept up jottings from his floor,
Bound the Enneads, and leapt further
Stretching to St Augustine, who dismissed [199]
'God's footstool' (the created world) as
Scant deserving passing prayer.

As raptors feasting on a carcass
Select soft offal, juicy liver,
Before they pick at joints or bone;
So the later commentator
Took good ideas, selected some,
Twisted the message until it served
An unintended dismal purpose,
A perch for new authority.

Involution

My move, I think, to amputate
Conclusions inappropriate.
Plotinus's era was more remote
From Plato's than the monarch on the current throne
Is distant from the Virgin Queen.

The intellects which build our empire
Go squirrel-wise from branch to branch,
Testing weight, assessing limits,
Narrowing odds on perilous guess.
Newton could take Kepler's pillars [200]
To place his new carved pediment
Einstein could float free from aether [201]
To invert and tilt the Universe.

The pedant intellect proceeds
By incremental building blocks,
'This hypothesis or that?
This equation seems to fit…'
Each sufficing (seemingly)
In retrospect is arbitrary.
The pyramid that then results
Concludes the structure at a point

(Did existence have a start?
Does God exist or does he not?)

Contemplatives probe differently
When they from swoon or trance arise.
They tend like bees to foster hives,
Ascetic is austere but sweet:
The man develops, not the thought.

The perfume that hangs upon the air
(The flower heads of former years)
Remains beyond the petals dropped.
The sirocco blows from Socrates
Across the dunes of centuries…

Canto the Fifth: The Dark Ages, Monastries and Muslims

The Bedouin in Averroes
(Serene upon his camel's back)
Breathes jasmine, rosemary and thyme
From rapture in Hellenic hills.

A script for such is comforter,
A trumpet call, a lasting post;
Another 'someone' made the journey…
Confirming the ephemeral
Elusive, fading certainties.

The soul's recurrent champions
Are dreamers, who in solitude
Find one another across years.
Each taper that quite sudden flares
Attracts the corresponding sparks,
Acolytes and scrolls arrive
To stoke the fires of new hearths.

Herein lies a cogent query:
If Plotinus echoes Plato
Or links Socrates with Averroës
Why does science turn its cheek
From such repeat experiment?

You have jumped the centuries,
Your knight does not respect his place.
You rude usurp my argument…
We cannot linger any longer,
Let's head towards Byzantium.

Arabs now were passed the baton,
Persecution herded them
To Gondisapur, where Nestorians [202]
Into Syriac rendered Greek.
The Bukht-Yisheu's favoured sons
Turned Syriac to Arabic.
Baghdad was now an outpost Athens

Involution

On a hotter dustier front…
Jabir, Rhazes anchored alchemy [203]
(Beyond transmutation's poor retort)
Hippocrates in them re-surfaced
For diseases and their cures:
Smallpox, measles and contagion…
(A text book for six hundred years)
In Avicenna's closing Canon.

Further East, trade with India
Infected them with algebra;
Abstract postulates gained sway,
Al Khwarizmi grasped that place [204]
Altered values either way,
Transposition, restoration…
Geometry was reconfigured
In a new economy.

Alhazen prefigured Newton [205]
On rainbows, haloes, burning spheres…
Rather than projecting vision
The eye received, refracted light.
He grasped the lens was there to focus…
Concentration followed suit.
When Cordoba was Arab captured
It built another library.
(Moors in those hot and halcyon days
Were very cool about the Jews)

Maimonides was Arab-Jewish, [206]
Talmudic scholar, translating the Torah
From Hebrew into Arabic.
Fleeing from Cordoba he found shelter
As court physician to Saladin…
(Instead of Gaza, Israel might annex
His perceptive, short and pithy
'Guide for the Perplexed')

Canto the Fifth: The Dark Ages, Monasteries and Muslims

Averroës has been mentioned,
Another Neo-Platonist…
The Universe was, for him, eternal,
Evolving (as in Darwinist),
Splinters from him pierced, infected
The Church's later 'heretics.'

Curtail your list, get to the point.

You are impatient and unfair…

You gallop low in the saddle of God
Lassoing some selected steers:
I play Border-collie dog
Rounding up entire flocks.
Your song has a repeating verse,
Mine builds across three thousand years.

You wait till you meet Gutenberg, [207]
His moveable stamping clattering press…
The hoi polloi will learn to read
Then you'll have your work cut out.

If I pay tribute to the Muslims
(Who get, these days, a jaundiced press)
It was for their crucial timely passion,
Taking parchments full of Greek
And burying them in Arabic.
There they incubate and grow
In Cordoba and Toledo,
Returned to Sicily, where first nurtured,
(Now orphaned by a foreign tongue)
A Tunisian, called Constantine [208]
(Murmurs Hail Marys; he inhabits
A monastic cell in Salerno)
Re-renders Arabic to Latin
Restoring much first lost in Rome.

Involution

The climate of creative thought
Seems an exiled shoal of fish
Recovering a spawning ground…
This sprint to Moorish Spain through Persia
Follows the filament waft of smoke
From beacons lit on the Grecian coast…
(The Church in Europe in the ascendant,
Free thinking now was all but chained)
The dark ages of a thousand years
Would need a new Olympic torch…
The Arabs kept it oiled and shaved
To light the common-rooms and quads…
Scholastics meet, and disputation
Founds the universities.

Grey and Black Friars first to High Table [209]
With God at sharp elbow, divine dispensation
To apportion the haunch of a rich education
Poor scholars receive a nourishing meal…
(Grosseteste ladled with slices of Bacon, [210]
Leonardo of Pisa, with Roland of Parma,
John of Peckham excused his tedious duties
At Canterbury, blessing pilgrims on mules…)
Quietly spooned from selected side dishes
Planets in orbit, explosives and optics…
Bacon devised a theory of flight…
Carefully concealed between linen napkins
Morsels Aquinas had not expressly ruled out.

Any science free of Church prohibition
Still crept with hoary hounds on the floor
For scraps that might fall from the Master's table…
The signs of the heretic, detection of witches…
Nicolas of Cusa's improper conjectures…[211]
Who'd been appointed to sit on what chair…
Starved of meat from the pungent kitchens
Science will finally gnaw through the legs.

Canto the Fifth: The Dark Ages, Monastries and Muslims

The slow swing of the pendulum takes a wide arc
Through the early centuries.
Man was looking all about him,
Making sense of everything.
Science now will narrow focus,
Oscillation will increase,
Drying out on Petri dishes
By separating disciplines.
(Mind retracing evolution
In reverse chronology.)

Arrêtez, Je vous en prie…
I demand a moment's pause…

You take your linear travelogue
Through catalogues of single men:
You forget this involution of all minds
Is with all of mind behind…
Or leading as a current does
Below the keel, or in the wind.
All creation is involved;
The pangs that birthed the anthropoid
Left stretch marks in the atmosphere.

Man, myopic, is self-centred;
He records his proud ascent
Above, beyond the bounteous planet
That gave him birth and tolerates
His pillage and indifference.
He shares his lineage with ants,
With humming globe, a single tree…
He calls them all his great estate,
Forgets that most preceded him.

Your history repeats this fault by keeping to the single scent;
You chase off theist dogmatism with the hounds of shackled men.
In your haste you are neglectful…
Science is far from solo voice…

Involution

It falls to me to readdress imbalance in this narrative…

Crucifixion cast long shadows over mortal man's deserts;
Rome was ever held accountable (fastidious washing did not help)
What you called 'a scrape of dripping' was redemption from that guilt.
The Christian churches sought atonement with penitence and some largesse,
Commissioned painters, sculptors, masons; bestowed exalted monuments.

Have regard for Celtic Christians; [212]
Like thrift they cling to sea sprayed rock
On Scotland's far-flung hostile islands
Iona and Lindisfarne…
St Columba croons the liturgy
To the silent bobbing seals, moved
They comfort him by weeping, all
Wait for converts who can speak.

In bleak monastic stone scriptorium
His monks transcribe the sacred texts,
Then bind the precious parchment word
In fine wrought leather, golden filigree
Set with gems cut cabochon.

Far away in Frankish forests
Charlemagne, the soldier statesman,
Drives the Saracens from the Loire,
Harries Avers, Moors, the infidel
Beyond the Danube and the Elbe.
Establishes three hundred countries
(The embryonic Holy-Roman)
Under a papal cross and cope.
Prefers Byzantine shape for churches,
Hires Anglo-Saxon priests.

At Fulda is new script adopted, [213]
The Caroline Miniscule
To render all the classic scrolls
Closer to original. At St Gall and Reichenau

Canto the Fifth: The Dark Ages, Monastries and Muslims

Abbeys thrive, all self sufficient,
Each a minor city state; coining, printing,
Artists' markets, silver, straw…
Episcopal palace is almost Islamic
In laws for both belief and practice, but
Erigena still reads Greek.

Avoid myopia: science is not here defined
But the preservation of the texts
Ensures the primacy of books
(More precious than the crucifix)
Charles himself, well past forty,
Tired of fighting, learns to read.
The first millennium, the darkest age
Was lit by abbeys, all enlightened,
Roman in name, but not in creed.

Later spread the daughter priories,
Under dispensation ruled. Kings
Consulted Cluny's Abbots
When they could be spared from prayer. [214]
Artists were engaged to offer praise in filigree gold and stone…
Resurrected Grecian fluting discreetly in the trios standing
At the great west door of Chartres. [215]

So the torch of the enlightened
Was protected
In these scattered colonies.
Crusader Dukes paid handsomely
For monks remaining on their knees.
The Virgin, now with sinuous curves,
Refined all members of her sex…
Troubadours plaintive, sang their lays
For widow, dowager and maid
To whom all sacrifice was owed…
The longing was itself the tribute,
Nobility had taken root.

Involution

Power and administration,
Humane law and giving alms
Matured within this federation…
Pilgrims, tolerance, tongues and news…
Nunneries and education
Spread throughout this holy world.
If wisdom gutters in the candle
Set behind a holy stoop. Give it due;
The draft admitted from the door
(In its breath the small flame withers)
Brings flaming torches, plague and pillage
Rape of virgins, aged mothers, all of savage sacrilege.

Had the godly then lacked faith
With what else was courage kept?

We now proceed to the era of Dante
(Sighing on that parapet)
Who classified the ranks of evil,
(A metaphor inferno earth)
Remember that even serene Virgil
Did not pass to Paradise.
You should follow his example;
Poets, words, and intellect
May explore the lower reaches…
At salvation's door they cease.

Let me weave an ivy tendril
Through the hair of Beatrice…
Painters give a clearer picture
Of man's conception of himself.

The prehistoric hunter-gatherer
Drew real cattle in red ochre,
Detailed movement, dust and muscle
Startled, galloping in herds…
(Food was central, close observed)
The men, with bows and arrows chasing,

Canto the Fifth: The Dark Ages, Monastries and Muslims

Are cursory, no more than sticks.

The Greeks show man in every posture;
Fastening a sandal, a charioteer,
Women in high-waisted dresses
Gossiped, drank, and dressed their hair.
Man was fully three dimensional,
Sufficient in each modest gesture
Pouring oil, or plucking lute.

Rome made painting decoration
For new pomp and circumstance
In country atrium and loggia
Naturalistic figures sported,
Frescoed gardens covered walls…
Sculpture no longer celebrated
The daily round, the common task…
Emperors ranked all approaches
(Tiberius disdainfully laconic
Resplendent in his armoury,
An orator brooks all interruption
Imperious with upraised arm…)
Individuals captured power
With hawk sharp eye, high bridged nose,
Rome had conquered, and it shows.

For Christian Church all art surrendered,
Iconography controlled.
Flat, symbolic, mostly message,
The messenger unnamed, ignored.
Fishes swim in cerulean heavens…
Vines wax strong beneath His feet…
White stags nuzzle near the Virgin…
Priests beseech with open hands…
Stone mosaics in Byzantium,
Fields of marble, haloes gold,
Saints and pilgrims, adoring bishops
In frescoed churches, miniatures…

Involution

Images considered pagan,
Natural man now disappears…
Christian crossed with eastern mystic
Hangs agony in wood or stone.

See, I now return you your period
(La peinture libre, a disparue!)
With 'book of hours' nun or widow
Went pacing out her daily prayers.
Monks illumine opening pages
With bestiaries, and small grotesques.
Minute Psalters so enlivened
Long services for choristers.
The Word sublime, immaculate, draped
In celibate calligraphy.
Gothic emphasized the vertical
The sheep looked up and cricked their necks.

We have almost reached Giotto [216]
He'll recast space, recalling weight
In well fed clerics, often standing
Several inches above ground…
(Holiness was natural helium
Duccio floated slabs of stone)
Perspective shouted for attention,
Men were flat-ironed, most in profile,
Movement stirred in draperies…
Obeying gravitation's summons
For women weeping on their knees.

Please remember…

The world is all en-folded mind.
The yeast of any forward thinker
Leavens the whole loaf entire.
The passive dough's receptive state
(Chemistry or temperature)
Consents to rise or sulks inert…

Canto the Fifth: The Dark Ages, Monastries and Muslims

All thoughts un-manifest are there;
(Hitler's unfulfilled intentions…
Marx's glum prognostications…
Leonardo's flying machines…
Some nudes of Michelangelo…
Beethoven's rejected themes)
Continue 'in potentia'
No matrix or rich sponsor then?
Then all remain to be renewed.

A melody is carved from silence,
From a flat wall a painting leaps…
Its surface will accept a fresco
If well prepared, and light consents;
So ambient consciousness permits
This painting style, this string quartet.
If Mozart perched upon a bedstead
(As did the Fiend for Tartini) ²¹⁷
Dictating a new requiem…
The only thing that would be buried?
The reputation of the dreamer
Naïvely rushing to inscribe.

For Greeks, the natural world was total
And man, his thought, was, to it, central:
The mediaeval theologian
Saw nothing but the path to God.
Man cannot lose the Grecian talent
Just because its value alters,
In the climate of a time
It abates, the work's remaindered…
Junk DNA is fossil thought.

Before I get transported
By bugle blast of Renaissance
I'll hand you back to my opponent
(The pedant nose to logic's print)
She ploughs her steady argument

Involution

Through builders who will crenellate
Pigeon loft, cottage ornée, and
Every thrusting 'point de vue'.

I flit among anonymous artists
Whose instincts clothe the atmosphere;
The narrow nave of matins whispered
Before the sun shakes off the dark
On cold chancel stones that welcome
Chaste and shivering bony knees…
The bibulous wake of fêtes champêtres…
(Watteau puts his maids in petticoats, ²¹⁸
Satin slippers, finery… Are they the subject or excuse
To show his skill with feathered trees?)
The oyster shells that they discarded,
The tattered tents, the trampled grass,
Give man's recovery of his journey,
Less exact but more complete.

Your move I think, and you're in check.

Triumphalism unbecoming
(Perhaps a little premature)
The pedant in laboratory
Fills rows of test tubes, stains his slides,
Squints down microscopes for hours…
Your artists had their young apprentice
With his pestle, grinding pigments,
Stretching canvas, mixing size…
Let loose later on some drafting…
'An arch, perhaps, in that corner
Rather crumbling… and what about
A hound behind that stockinged leg?
All animals add liveliness…I know…
A whippet like the patron's dog?'

Canto the Fifth: The Dark Ages, Monastries and Muslims

All building has its jobbing masons,
Its scaffolding and mortar mix;
You choose the company of artists,
I couch with master architects
Who coffer ceilings, place a fountain,
Wind a broad and gracious stair
In constructions of the mind
Before a single brick is laid.

Stale mate I think; you cannot castle,
I cannot win with 'either-or'
The 'both-and' in each case stays silent…
(The mind behind the action's impulse)
This board's a perfect metaphor.

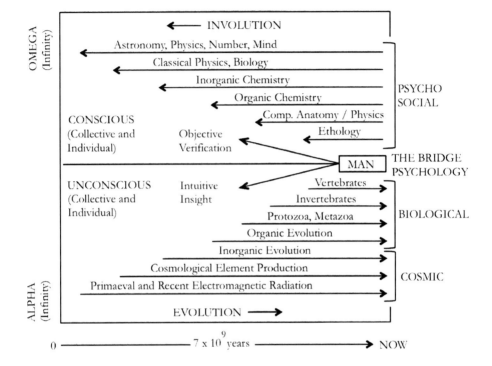

FIG.5. THE EVOLUTION OF SCIENCE

Scientific Disciplines emerge in reverse chronology, as memory is recovered. From the 'holistic' science of universals; number, cosmology and the Structure of Matter (Greece) through the disparate disciplines of Physics and Chemistry, later Biology and organic Chemistry, Physiology and last of all Psychology and Ethology, as man re-approaches the nature of himself and his bonds with all creation. All continue throughout.

Canto the Sixth: The Renaissance and Liberty

Look, are those not single blades of grass?
Our traverse through the deserts soon will cease…
Take heart, and now pray tell
How do you envisage Renaissance?
The babe delivered to the Virgin Queen
Who grows quite pale and middle-aged—
(Too many suitors mooring high-riding barques,
Too many intrigues by earls with dark looks,
Her hair a cascade, Rossetti red,
Her geisha face already embalmed)

Is not science's bairn pulling her skirt
A clamorous welcome, exuberant change?
From the rheumatic edicts of King Theos?
So palsied with his hand upon a cane
He scarce remembers how he came to reign.

I see it as a single peony, blood red,
Coiled as a bead upon a slender branch…
Sparse leaves are lances stretched toward the light;
Tight shut it stays until the bushy growth
Of foliage below will bear its weight
When fully ope'd. It will amaze
With myriad petals of enfolded silk,
Each fritillate, each crumpled like a waking babe
Scarce touchable, yet so enamoured of the liberating sun
It risks a fall through excess worshipping.

Involution

Yet, whence this billowing smoke, this smell of tar?
I smell a charnel house of violent death.
How are these hectares to horizon stripped?
What is that blistering of faggot logs?
Have we strayed from our path? Are we now lost?
I thought Inferno we had journeyed through
This hell is unexpected... is it misplaced?

Your peony is bloody to some point:
Giordano Bruno burns upon his stake [219]
Setting alight a landscape newly cleared
Of Vatican vine, the smothered close control
Of thought. Hitherto no adventure could find root...
All phoenixes need ash from which to rise
Else sodden history weights a laboured flight.

Bruno was impetuous, half in love...
Mnemonic trickery of rhyme...as in
B for believe nothing until you think it out
R for rotation of the earth, revolution round the sun
U for universe, uncreated and unsought
N for nothing born and nothing dies
O for omniscient, obvious to one who thinks.

He does not burn for rhyme but message implicate;
God finds no rule, no role in this man's thought.
An infinite universe lacks both end and start,
The heavens seemingly go on and on...
The barbed barrier of fixed stars dissolves like salt...
Man's mind may now be gainfully employed
In reconstruction of a rational God.
The Inquisition over-played the Church's hand
(Oppression does not sup with Prince of Peace)
In petty conflagration is its power waste.

A new crucifixion of a plain-speaking man
By those who had good reason to reflect
That martyrdom is paraffin for flame...

Canto the Sixth: The Renaissance and Liberty

But see, the landscape is a virgin plain,
Man's mind restored and all his own…
(Their silver soul, enwrapped, they now assign
To tarnish with a man in a soutane;
Surety for mass and marriages
Until at death it will, by gatekeeper, be cashed)

Let us now make sense of what we find—
Where shall we start?

Now I have surveyed the coming centuries
From the plateau of mediaeval lethargy
I am amazed. A hive of industry, a termite mound,
A frenzied breeding ground emerges as I watch.
Raising up a pinnacle of muddy escalation.
All busy in their independent task
To build the scientific citadel
From huts haphazard, built in what-you-will.

I am a Duke Wellington, like him perplexed [220]
By a survey (not unlike his coast of France):
A flotilla of sail at his broad back,
A narrow telescope to eye…
Is that serene beach well defended?
Are those slow cows? Or canon mounted?
That headland out behind the squall?
That cove with breaking waves on rocks?
Is there perspective via sky?
Could sail transmute to parachute?

I must adopt a deeper stealth
To find a vantage point for sipping sense
From such excess. The printed word
Finds porous borders. Chronology must now submit
(Thinking men think outside time) Not single spies [221]
But in battalions. Did such explosive new endeavour
Anticipate an end to chance?
Will I deflate this swelling bread
By nibbling currants of such busyness?

133

Involution

May I suggest we gather up those smoking bones?
Find anonymity in some quiet parish church
In misty Oxfordshire perhaps, where once he came…
And where his thought joins Newton's, in some years anon…?
Above the bones lay out two centuries abreast,
For they are joined in mental marriages,
As is that noble Knight, his visor down,   [222]
His Lady in her kirtle parallel; extend
Their tomb to Europe's boundaries,
(Here I do curb your trend to over-egg)

Their narrow breasts must rest in Tuscany
Aswarm with love of every subtle sort…
(Artists have revived anatomy; Pollaiuolo   [223]
Draws muscles with insane delight; Masaccio
Explores perspective for the start of space; da Vinci
Plays at catapult for Ghibelline or Guelph   [224]
Or keeps vigil at a deathbed to dissect…)
Let their buttocks rest in Spain, suppress
The still strident Salamancan Jesuit…
Beneath their knees in Portugal,
Diaz, da Gama and Magellan set their sail   [225]
(Cartography still needs a ready port).

Britain, I believe, must claim the feet
Dampened in the fens and willow marsh,
Where Newton must be surely granted
All blueprints for the rising edifice.
You may disperse those seven maids in farthingales,
Those brother pages that are sacrificed for wars
Or service to some state, to Scandinavia,
The lesser province in Germania. This bier
Is consecrated to the rising progeny of man
Who takes his parentage in history.
We shall not permit chronology to suffocate
When all is patterned to collective mastery…
Place has more influence than time

Canto the Sixth: The Renaissance and Liberty

On men, in what they choose to understand
Or how they spread their rediscovered world.

That being done, let us take wing
In a basket swung beneath a striped balloon.
We can drift across two centuries of time;
View countries at a glance in synchrony,
Pick out the teams of men, the drays that strain,
The signals that are sent in paper darts…
For the belly of this fertile time swells on its own.

Its impetus has reached a birthing point: The fisherman
(Who sits contemplative…alone,
His line floated to dark margins of a pool)
Can only land the fish that leaps.
The future is that swerving shoal
Erupting from the surface of the past
To take his clothéd hook of searching thought.
Each worker seems oblivious
To others; each a cell, enclosed within a sorting skull.
But there are patterns to discern, and countries' differences
Choose language and interpreters prepared. As builders will
Find raw materials quite close to home.

See, I have spared you any need
To weave your argument too rigidly
In either time or place. Space-time is both
How we travel, and how all mind's dispersed…

Think no further than our puny sceptred isle
With gneiss rocks half as old as earth, [226]
Battered by tectonic plate collision,
Volcanoes, glaciers, deserts, flood,
Sheered and upended, like an oolitic laundry press [227]
Exposing sediments, or plasma, granite folds apart…
Within the compass of an evening stroll.
More geology beneath the boot
Than through the whole continent of France.

Involution

(The French have food, prefer domestic warmth,
Chemistry burbles like a seasoned pot-au-feu…)

If the wanderer was not inclined to watch the ground
He might see cottages in flint or plaster lathe:
Green sandstone on the edge of Salisbury Plain
Turns gold near Ham Hill, white in the terraces of Bath,
Bruised red in Devon's ferrous clay, or marble grey in Portland Bill…
A week's walk would fertilize Hutton or a Lyell… [228]
Their library from Scafell Pike to Dover's chalk.

(Move on. There's coal in Midland mine, which spawned
 Brass bands in Yorkshire, singing in the valleys of North Wales;
The British Raj was built on railways, bridges, steel
Mills which split the emerald isle, in North and South.
The French army fought the Duke at Waterloo
In greatcoats clattered out in Lancashire.
Britain attributes influence to stalwart men
Of discipline, rigour, and stiff upper lips…
Instead Geology's impervious hand of cards
Fell open here, just luckier than most)

The history of the Earth itself, from sediment or fossil life
Shaped nations and their character. Greenwich set the British clock [229]
For a science of caveat and cautious lists…
(Wallace through travel was touched by the sun; [230]
Evolution came to him at a glance…)
Darwin made years of copious notes,
Followed by rows of greenhouse plants,
Amending, cross checking, endless delays…
Until Wallace and death were drawing abreast…
They nearly managed to pip at the post
Scooping the credit for diligent work!

Time may not be relevant nor chronology.
Our subject is the synchrony
Of universal Renaissance…

Canto the Sixth: The Renaissance and Liberty

Mine is, only momentarily,
The influence of clime and place.
I thought you would wish to emphasize,
The organic birth of national thought—
The French are comprehensive thinkers,
(Start with Descartes, to Voltaire move on… [231]
Eclectic in topic, exhaustive in depth)
Englishmen only trust their own work.
Newton's 'Principia' was far from a pamphlet,
Darwin's 'Origin' left little out.
The current crop is just the same;
Six books (and counting) on the greed of the gene. [232]

I believe the country of most recent cloak
Weaves its tapestry of silver thread…
(Or homespun Arran cardigan)
Like-mindedness congregates with others
Comfortable in certain chairs…
The Oxford don is not from Cambridge,
(Plato glares at Stagirite
The great divide will still compete)
Did Geology's slow pace enrapture
All succeeding layered thought?
For the nation who, at Greenwich,
Set the international clock.
No interruptions by invasions,
They could correct, perfect, as time would wait …

You have a point: you were conceived on safari
(A tarpaulin under a Baobab tree)
Small wonder that you shoot from the hip…
Do not your cousins all eat in the saddle and carry
Biltong in a shoulder bag? [233]
Not merely 'bleu' but still on the move.
Europeans prefer ideas well stewed,
Peppered with authority,
Puréed and spread very thin on toast…
Washed down with a glass of chilled Chablis.

Involution

Your thought is as wild as a set of bared teeth
Caught in a snare; A swarm of hornets on the loose…
Did you advise our reading companions to bring
Leather gloves and a bee-keepers hat?

That's rich. It was you took me up the high colline
To offer the blue haze of history…
Did you not say 'Eat what you will?
Nothing's forbidden, now the apples are gone?'
If I wandered unguided in that forest alone
You slithered away like the snake that you are.

Tut, tut, you ought to take a tease,
I can only answer the questions you ask.

New orthodoxy now deems incorrect
Any mention of bias to national thought…
If Germany produced more music than wurst,
Laid tables of crisp white philosophies…
England language: playwright and poet,
The French stayed at home and thought very deeply,
Italians build dam and bridge, drive stanchions
Into the sucking sea (They were Romans once and perhaps still are)
(La Scala Milan is their new Coliseum,
Divas still hissed or lionized…)
Could it be the legacy of great men's absorptions
Bent their separate spectra of inspiration
To focus their countrymen's continuing passion
For the children that succeeded them?

A deeper instance of 'climate of thought'?
What flourishes is already seeded
In both the mind that connects, and historical tilth…
The Renaissance found ready thinkers
Throughout Europe but each language adopted
Had 'inflexion', a regional 'twang' or 'cast',
As though the concert of wholly human endeavour
Was a choir to blend universal song?

Canto the Sixth: The Renaissance and Liberty

The praise of creation through the creators?
That seems to me to make perfect sense.
Now that is established we'll enjoy the detail;
The quinces, cherries, old and new apples
Bursting out on the single tree. They preserve
The balanced profile, filling in the girth expanding
Through men universal with syntheses.

Before you draw the curtain aside
On a stage divided by balsa wood walls
May I review what we seek to show?
That all of thought is thrusting now.
Renaissance recovers the innocent eye
Of the Greek, his world, abundant joy,
Adapts it anew, at a distant remove
For a world of trade, and the printed word.
The Roman trireme that served the Aegean [234]
For the Atlantic gale needs wider girth…
Man has expanded across the globe,
Columbus discovered a glowing horizon,
Da Gama rounded the Cape of Good Hope…
From classical Greece, Europe derives
Values of liberty for all but the slave.

Aristotle, pickled in formalin,
(Aquinas' strictures had bled him dry)
Is not now summarily pulled from his plinth
But crumbled, made friable for a new mix…
Descartes, the father of rationality
Began with his 'cogito ergo sum'
Platonic in emphasis; (thinking is proof…
Aristotle remains in the subjects of thought)
Rational man can decipher it all
By shaving the process of thinking itself…
Mathematics re-marries the natural world
In the mind of the thinker splicing both.

Involution

So the Greeks we reviewed at the start of time
Remain like good compost, well embedded;
The oscillations alter between thinking and seeing...
Science supposes solutions exist
If conjecture is followed by its proof.
Tools are aids in this pursuit,
Lenses and mirrors expand the sight
To new stars and clouds of enveloping gas...

The Church retains value not often acknowledged,
Its domination and power still needing defeat
(The oyster needs grit to lay down a pearl)
Which proof mathematics and optics provide.
The river of thought that it had suppressed
Gained in pressure throughout subterranean years...
When the crust of obedience had sufficiently thinned
The artesian force found diverse men
Unafraid to ask questions or leak simple doubt...
Finally that waterspout blew as a whale,
Exhaling at last!

Before they spring forth let us drift over Paris
Where Pascal and Descartes are summoning Plato
In their tendency to enlarge the minute...
The small fly buzzing near the ceiling? [235]
How would you define its position exactly?
Two lines intersect at fixed angles to others
(Stationary for split seconds of time)
Cartesian co-ordinates describe a curve...
Algebra now joins hands with geometry,
Equations relate number to form
(Graphs have plagued us ever since)
Re-uniting the inner and outer worlds...
For the latter Pascal sent his brother a climbing [236]
Up five thousand feet of the Puy de Dôme...
Tiens! Barometric pressure changes with height
Confirming his thoughts on resistance and vacuum...
The brother descends: now air has weight.

Canto the Sixth: The Renaissance and Liberty

Aristotle was also slowly crumbling
In Britain, where Gilbert, the Court Physician, [237]
Absorbed by magnetism, loadstone and compass,
(Observing capricious signals to north)
Only explained by the poles of the earth
(Like his Monarch) itself a force of influence…
So too the stars, no longer fixed,
But now the centres of planetary systems…
Bruno, long burnt, had said as much.

The uniform systems of thought were dissolving
In the fluid of new enquiry.
(Yet now begins the binding together
Of varied phenomena strapped by new laws
On the head of the rule of Ars mathematica,
That weaves by gathering varied brushwood
For the fire that reduces all to coal…
Science's priesthood will become tyrannical;
For the moment they splash in a paddling pool)
The heavens were also shown a-changing,
Man's mind was all he had for certain
(As long as his mouth was under control) even so
The Church slammed the lid of Pandora's Box
On the hands of the man we now introduce.

Galileo Galilei. [238]
Pause. Stop. Repeat.
Galileo Galilei…Galileo Galilei…
Is it not an angelus bell from a Tuscan hill?
That thin and penetrating timbre
Riding high above the heavy heat
Indifferent to the tethered goat,
The still reaper that stands with his hat in his hand.
The habit as constant as rising sun
Noticed only when it fails to come…
The piazza deserted; pigeons throat in a belfry, subdued.
On Fiesole hill he sits at a table, the shutters serving
Slices of light. Imprisonment is a blanket now

Involution

(A warm enclosure for dwindling life)
The fertile uplands of his thought are liberty enough.
His eyes are dimmed, almost totally blind;
That matters less, his work is done.
Milton came but yesterday… [239]
Poets are blind, even when they see…
Both recall a Paradise Lost
(The faultless order of man's estate)
The lunar surface is now pock-marked
With seas and lakes and mountain range.
The Virgin orb that drove men mad
With love in its ascendant phase
Has a face diseased by looking too close;
This apple has a sour rind.

For the thinker who would fain retain
His prior concept of the good?

'Il Saggiatore' aptly stood
Where mind and matter intersect
New links perceiving wider breadth
Expands the world beneath the foot.
The innocent eye that looks anew
(Or takes its stance across a void)
Between the fertile disciplines
(Languages of equal force)
Transforms in an instant what it views:
The world of yesterday is changed.

A colossus equally at home
In medicine, music, mathematics,
'Disegno' (perspective, chiaroscuro)
All gave to newly binding law
Harmonious geometries.
Dynamics was the flow within
His mind's high circling thermal rings
That perceived, hawk like, that movement below
Compelled dynamic narratives.

Canto the Sixth: The Renaissance and Liberty

Mechanics was the law of change;
Rest no longer eagerly sought…
The moon had aged, vicariously
From Dante's smooth 'eternal pearl'…
Stars in infinite succession were
The substance of the Milky Way.
Spots on the sun had influence…
(Out went much of Tycho Brahe)
Venus had phases; all revealed
A Universe in flux, suspense…
The good book of the changing world
Was in vectors and equations spelled.
Pythagoras breathed fresh life, arose…
Aristotle's hierarchies quietly died…However

His careful methods were retained
In tools this lucid man devised.
The 'starry messenger' loomed into view
Through his telescopic eye.
The compound microscope discerned
The miracles of insect legs,
Their fine filaments and mandibles,
(Their compound eyes had got there first)
The expanding air thermometer
Kept watch upon the feverish ague…
The military compass could assist
The estimates of acreage…
Jupiter's moons would now provide
The standards for a constant clock.
The world no footstool but a spring
To quench and elevate the mind.

None incidental, all cohered
(The mind the measure of the world)
This 'father of science's' appetite,
(Eclectic but full disciplined)
Made of the simplest small observance
Universal mastery.

Involution

He comprehended light and sound
Had frequencies: relativity was implied
In his law of constant speed.
Newton and Einstein were both launched
Like ships from Galileo's wharf.
Since Greece no equal polymath
Melted matter (like good butter) into skillet of the mind.

Genius leads collective travel
Through all embracing syntheses.
Galileo straddled the perfect pivot
That checked mind against the turning world.
Poor Kepler at the self same time [240]
Preferred perfection; tried to steer
His observations to conform
With pre-conceptions of the pure.
Plato's gaze still plagued his sleep
(An eyeglass to ignite a fire)
Seeking burning harmonies
In the music of the spheres,
Platonic bodies intricate
Shaped heaven, as would five hands a glove.

Yet he was right about the tides
As minions, in their rise and fall;
Pytheas had first observed
The relations of the tides and moon
Was like a stable, well honed marriage
That bickers, never goes to war.

Galileo argued: é tutti sciocchezze' [241]
For him the tides sloshed back and forth, dragged
By earth's rotation round itself.
Like Archimedes in his bath
The tangible was more than half.

Genius itself will vary the mix
Between the thought and the manifest.

Canto the Sixth: The Renaissance and Liberty

The building of collective truth
Is balanced on the need for both.

Kepler was in essence dreamer;
Circular was perfect motion…
His disappointment in ellipse
Was mitigated when he found
A new perfection in its speed
That traced out equal arcs of space
In equal times: The perfect now expands its stage
Towards a stately polonaise (motion is its mental trace)

The seeker who improved his sight
With lenses ground and mirrors curved
Was struck down blind in later life
To dwell entirely in his mind.

The dreamer, his counter, dispatched to Prague
To peruse the scribbled sheets of Brahe…
Where Ptolemaic figures scribed
The comets' regular returns
Through historic and observed accounts…
Precessions of the equinox…
Mars' orbit with eight minutes error
Reset his mind on solid ground.
To reconcile what he conceived
(Circular orbits of constant speed)
With tugs-of-war and influence
(The force of the sun diminishing
The further away the planets spun)
Physics now climbed the skies…adding
Dynamics of force to the crystalline.

Please note well this is the pattern
Throughout creation's slow advance
Opposites are reconciled—
Galileo fixed the earth bound laws
The heavens by Kepler were aligned

Involution

Mind and matter interact
To widen laws, but reduce
The separation twixt the two.
So, equally with time and space…
Until with relativity we get
Space-time; Galileo stretched towards that light…
Einstein grasped it with both hands…
Like Eros working all the strings…
On space-time hangs the puppet dance
(Hidden behind mc squared)
Manipulating incidents.

Do you imply that minds are shaped
By instances of trivial chance? So
Corrections alternate through lives
To better master or dilute
Man's path towards a larger God?

(Accompany me if you will…
Let's drift above dark cypress chalks;
Some questions are better sketched and drawn
In privacy and off this stage.

Trivial does not make much sense
In symphonic consciousness…
A curvature returns, vibrates…
Weaving through both time and space…
Relationships are intricate.
A piccolo will penetrate
The deep ocean of a unison bass.

We have correctly emphasized
That 'act' throughout is fertile seed. Creation
Is birthed in every thought, but act
Converts to permanence.
Men are tried for criminal act, but
In the future criminal thought
Will be held to scrutiny…

Canto the Sixth: The Renaissance and Liberty

(For thought lives on until it finds
The actor to enact intent)
The circus of the natural world
(Round which man cracks his heedless whip)
Is not mastered yet by man at all…
All creation seeks a shepherd
To pipe, from rock, a quickening reel.

That hungry boy in Tübingen,
Kepler, the frail, was quite alone.
His searches of the starry dome
Were for the Universe within…
Mysterium cosmographicum,
Harmonia Mundi, de vero Anno,
Perfection rings in every term.
From music of the planetary spheres
To symmetry in flakes of snow.
He ended with his 'Somnium'
This work 'the dream' that closed his eyes.

The Tuscan, Galileo, in benign Arcentri grew,
Affluence in endless sun; the music of domestic lute,
Mathematics was his father's tongue.
Fair Florence's exuberance
Gave painting, building, medicine
Explosive wide equivalence
In which he walked as through a garden
That blossomed as he looked at it.
His privation was denial
Of his due entitlement.
His work was banned, his speech curtailed,
Followed by his house-arrest.
Finally in his old age (perhaps prepared for modest death)
Was sacrificed his perfect sight.

So you are setting out to show
Hereby, such men are puppets too?

Involution

Puppets? No. But instruments,
Honed by lives and natural talent
To fulfil the total mind's delight
In shunting home to origins...
(It's easier to re-organize
A cupboard not yet overstocked;
When there are spaces in-between
The shelves accommodate new facts)

Polymath does characterize
The early breed of genius;
A little of everything enriched
Wide sweeping laws they feasted on.
(Later we will have to come
To narrower men of quantum thought
Where language is itself obscure
To any but the cogniscient)

So tell me, plainly if you can,
What is the thought that finds such men?

Consciousness, like pure white light
Is omnipresent everywhere:
Its spectrum is a thousand bands
Refracted through a focussed lens...
(Creation as a whole partakes
In sending hues of subtle kinds)
Each band diverges or will seek
The sympathetic string to sound...
An antenna, whose searching, reaches up,
Probes fallacies, until inflamed—
The unexplained miraculous,
The half-familiar, unexplored.
It palpates the dull conventional
(The orthodox compliant dark)
For the imagination that might spark
In love to form a natural bond.

Canto the Sixth: The Renaissance and Liberty

The mind in which it finds a home   242
Is often childlike or naïve, both sceptical and credulous,
Willing to dream the larger dream, lofty in its wild surmise;
Prodigy or innocent, but tethered to the relevant.
These are the minds…dissolved in curiosity,
The gaze is steady, unremitting
Finding what they seek to see…
They marry what is, to what might be.

Creation has its pregnancies,
Delivers intermittent gifts
(Sometimes twinned and synchronous;
New ideas sprout like plants
Bizarrely simultaneous
In distances of barren ground)   243
The intellect we call the 'mind'
Is but a brain that resonates
To music played outside the skull:
The piper pied that calls the tune
Leads on from deep, to higher ground.

Mind and matter will one day find
They are identical in kind…
This journey now is almost done;
We have re-approached those Greeks
The whole is where the story lies…
Shall we leave off pounding grains?
(Smashing matter underground)
To listen to the call of cells?
The hair that rises on the nape?
The stomach pit that speaks of dread?
Fiddlers' fingers that recall
At speeds beyond the rational?
Why characters in dreams appear,
Why they segregate and stay
Behind the ravelled sleeve of sleep?
Let consciousness take its rightful seat
The High Master of long patronage?

Involution

This war that now engulfs the globe,
The fury of the natural world,
The hatred, cataclysmic greed,
Jihad and terrorists with bombs…
Symptomatic of this re-approach
To universal consciousness?
That makes us all subsidiary…
(Neither author nor autonomous)
Is mankind unwilling to embrace
His dissolution and turn round?

Severance and separation,
Is the price expulsion paid
For knowledge of the 'otherness'
The mirror glass is still intact
Dividing us from all that is…

The pendulum now is almost still,
No metronome controls the tune;
The podium seems vacant, but
All clamour and would fain conduct;
The score mislaid, and not yet bowed, ²⁴⁴
The key itself is undefined…
The concert is now being tuned
By discordant soloists,
Trumpets blown are all their own…

Let us return to hopeful times.

I realize you will soon proceed
To spoon the queue of single men
Who, by degrees, in different ways, dissolve
The substance of the hard and fast
By laws of transmutation that will show
All matter is in constant flow.

But let us first return to Rome,
Now done with conquering routines… instead

Canto the Sixth: The Renaissance and Liberty

A cross-road in candescent light where history convenes
A Pope, at last, who nods to God. ²⁴⁵
His glory and his Lord's unite...
Two painter's paintings spell it out
What we, in words, approximate.

Come, take my hand, and we will climb
A perilous scaffold where one man
Lies cramped upon some wooden boards,
Paints 'creation' in the smoke of flares...
Cold he must be, in rigid attitude
In this locked tomb for four long years.
His eyes fixed on the work above
Must also take the measure far below...
The nude perspective tailored for the ground.

On heaven's vault he drafts anew
God ennobling and disposing, rousing man
Reluctant to abandon his stretched hand;
Already his disgrace is sketched
In 'expulsion' that will follow next.
Whole man is one, in contoured flesh,
His mind, its energy, its thrusting life,
Its sombre languor is expressed:
God and man are single, severed only
By the freedom's parting and its doubt.

This giant of the perfect form
Made man patrician, masterful, complete.
Perfection was man's natural state,
The figure gilded by high-minded muse
(Though he could carve or paint their attributes)
It was their God-like nature he adored.

Not far distant is its counterpart:
The 'School of Athens' is complete and all ²⁴⁶
This tale intends it illustrates...
Above a broad belvedere of steps,

Involution

Beneath receding arches, noble vaults,
The two Athenians stand in fullest light;
Plato with a finger points towards the sky,
Aristotle cradles heavy book against a thigh.
Spilling from them acolytes like cloaks
Side-clothe the steps but leave the centre swept.
On the left Platonists ponder, read, reflect, look up…
On the right, globe, compass and the astrolabe assist…
(They take the glances downward in this group)
Man here is in command, he built this space,
The marble steps, unoccupied, approach
The central disappearing focal point
This lies in light, between these two wise heads.

The classic world is here entwined
With reason's renaissance applied
To proportion and perspective space,
In which man's mind will alternate
Between his spirit and its carapace…
Until those coffered vaults are sprung
To over-arch a blending, heart with mind.

Here are individuals, young and old,
Gracious, aged, arrogant, at large;
Man is full himself, in his created space
(If he serves some other purpose, it's obtuse)
Raphael sees distinction in his groups
But each figure is absorbed and isolate.

Since soon I will be shunted off this stage,
Told art is not the topic of this work…
I must not fail to draw a final point
That here it is where art and science part.
One, at apogee, to celebrate the form,
The solidity of man (the detail all distinct) his fellow creatures
(Crouching hare, the patient mule, the distant landscape almost real)

However, science now begins to blur the lines
Of matter's varied shapes and changing forms
(Liquids become gases under heat;
Both are molecules in constant dance)
The Renaissance solidified the natural world,
Gave human form its gravitas and weight
But matter in itself begins to melt.

I am done. You have the floor. I could not depart
Without reminding you that art is history's truer analogue
That runs like print through Brighton rock.
Bite where you will…
Those titans are immortal,
As true today as when they worked,
The genius of science has a shorter span, for
He is cut-purse robbed by later men.

Your artists occupy a narrow cage
Within the close constraints
Of commissioned work and patronage…
'We need an altarpiece…
Perhaps a Trinity? … Or grim Descent?
The 'Healing of the Leper' would give space
For Signori who will be flattered and may well
Engage you for that convent being built…'
They bent old ruling limits to escape
The tedium of fables over-told:
To draw the thorns from impulses confined,
Audacity would startle with surprise.
(The intellect stayed subject to the heart)

Though often young— (Bernini was but three and twenty when
Julius engaged uncharted wild excess
To colonnade the Vatican's piazza,
And give ecstasy the promise of good sex)
They were all apprenticed
To their demanding master or a guild,
Familiar with the others in their field…

Involution

Techniques, tempera, etching, fresco...
Were detailed in the colours then at hand;
What else could they contribute but themselves?

Take Florence, liberated by its wealth
(Medici bankers almost a new Church)
For whom Botticelli was invited to supply   247
Pensive porn concealed in virtuous myth.
Lest thought facile or irreverent
Melancholy pervades, makes nudity now decent.
Even Judith, (triumphant with the head of Holofernes,
Looks as though she's lost her trilling linnet...)
Wind wraps soft curl around caressing finger,
Flutters drapery against a shapely bosom,
Blows out the cheeks of fat and pouting putti,
Licks up wavelets shaken in a spoon.
(Breeze and passive pensive maidens
Were lucratively Botticelli's thing)

I could expand: A similar obsession
In Dürer's roving and audacious eye,   248
Which served all fillets carefully deboned...
He stroked each hair upon a frightened creature...
Each wrinkle on a toothless listless hag...
The pitiless details of the lucid surface
Somehow cauterize the sympathetic heart.

I would much rather make my argument
Through painters, for they travel the same route:
Take ship for Venice with its carving out of light, enjoy
Giorgione's ample nudes, (below a growling sky)   249
Get serious with Titian's dark entombment,   250
Propose reasons for the tranquil Dutch interiors,
(Flemish winters, and the health of Jan van Eyck?)   251

Canto the Sixth: The Renaissance and Liberty

But why address the already long converted?
When science is in greater need of vision
Believing, as it does, it holds all truth.
Scientists were caged, not by commission,
Invisible are the bonds to its repute…
'Nullius in verba' was its sole prescription; 252
What *had been* found, determined its direction…
What *could be* proved alone would stand respected…
What *was* endorsed alone would move it on.
Now to have some hope of any new deflection
That journey must, in essence, be retraced.

Mankind's initial stupefaction
That sought his explanation in the stars, has swung
Between heaven's inaccessible perfection
And the mundane footstool of the fates.
From the clear intelligence of Athens,
That saw the mind itself as part and parcel,
(The lens refracting aspects of itself)
Then fractured by Athenian sub-division
(The soul entrusted to the strident Church)
Swings back to rational conjecture
Which through the mediation of two men
Unite heaven and earth, the human body…
All tooled upon the forecourt of the brain.

A carriage, in the sun of understanding—
Its inertia, weight and moving mass…
Its colours match the moods of changing seasons…
Its lines are shaped by wind and tide erosion…
Its speed is geared to annual succession…
Its fuel is waste as simple food recycled…
Its light is sunlight, absorbed and stored, converted…
It drives itself, and needs no steering hand.

Involution

This machine so perfect, self-sufficient,
(Which produced, by error, man, his probing mind)
Now draws head-shaking, marvelling mechanics
Armed with spanners, wrenches and enthusiasm
To take apart this fully automated engine,
Its plugs and pistons, converters, axles, shaft…
With a will all occupy their benches
Now confident there is no lurking 'ghost'.   [253]

I suspect I may now gracefully retire;
I see we now approach the cliffs of Dover,
The Snowdon altitude ahead of Isaac Newton
Whose heady laws will leave us feeling faint:
(No mystery left to cloak or hold belief,
The air soon robbed of any oxygen…
We'll subside exhausted later on a beach)

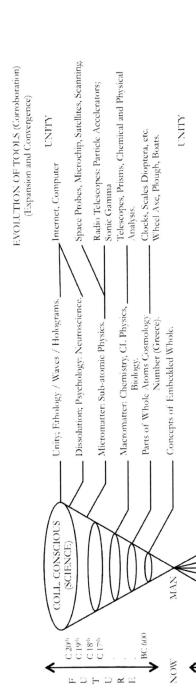

Fig 6. The development of scientific thought through the recovery of evolutionary memory

Canto the Seventh:
The Enlightenment and Rationality

The 'materia prima' is now a blameless lake [254]
Unruffled by caprices of the fates…
God has been banished from the modest hearth,
The vacant mind has hunger new unleashed
From comforts of the Church and of the past.
To penetrate new altitudes and depths
You may annotate, but keep the script compact,
Explosive thought will cover all with ink.

The face that launched a thousand derived laws
Has a nose that might have been designed
By Isambard; so cantilevered over bone, [255]
Escargot nostrils finely raked and shaped
As any aqueduct processing on to Rome.
Eyes of glowing coals that burn
Below a stalwart mantelpiece of marble.
His chin eccentric as a boulder swept [256]
From distant glaciers of melting time…
Full centred in the flowing cataracts of hair…
His gaze arrested in the moment's pause
To spear a leaping question never asked before…

Newton was a quart-jug midget child [257]
(Another orphaned by his father's death)
Put to the breast of a widow in a Manor
Somewhere in Lincolnshire, destined (if it lived)
To till the soil, play farmer if it would.
Undistinguished, except right obstinate,
Shunned Aristotle, nobly leather-bound, instead

Involution

Took Descartes and a candle up to bed.
'Hypotheses non fingo' was what he said
When he had turned the turning world upon its head
With hypo-theses that like pylons underpinned
All marching columns, categories of thought.

He picked out and polished Kepler's laws   [258]
Under the linen of a cold and narrow cot
When exiled to the country by the plague.
Extended Cartesian co-ordinates
To calculus (and later fought for it);   [259]
Derived equations that would fit
The falling apple (called The Flower of Kent) and
The spinnaker moon with passing orbit dipped,
Saluting gravitation, much perturbed
By the irritating flattened poles of planet earth.   [260]

Light was corpuscular and moved
Through aether that transmitted subtle force,
Allowing grains of light to penetrate and fall
As did an apple for the centre of the earth.
'Optiks' (through the splitting of white light)   [261]
Showed constancy of colour's spectrum; all
Bands refracted differently…but retained
Unique fingerprints that smeared the solid world.
Vision was encounter, lay between
The light that struck and rang the object's shape.

A 'force acting at a distance' without God
Smacked of alchemy (now in frenzied flight):
The last magician he was called by those
Suspicious of his wayward nonchalance…
(He accepted neither cloth nor narrow collar)   [262]
Not content to shape three centuries of science
He redesigned the coinage of the realm…
Took to disguises in low taverns near the Thames   [263]
To apprehend the 'coiners' he despised.
As Magistrate he met them in the dock,

Canto the Seventh: The Enlightenment and Rationality

(Played prosecuting counsel, judge and juror)
Dispatched the felons; some were hanged and quartered,
(Their thievery would suggest he'd made some error:
Outrageous to the Master of the Mint)

He died with neither wife nor living issue
(Too much mercury in his flesh and blood)
'Ah, that would reveal his latter-day obsession
With heresy and those Rosicrucians' [264]
He held that God was immanent throughout;
(The still focus of whatever was the system)
From where collective mass and movement, all
Defined a point to which the rest related…
(Here he fertilized and ploughed the field for Einstein
Who anon will find that focus, riding light)

Principia Mathematica now the new Good Book
To ignite a rational brave new world…
To set a thousand mills to spinning cloth…
A thousand furnaces to making steel…
Enlightenment would take its reasoned stance
To argue liberty across the human state;
In thought Voltaire applied his measured calm,
In other matters he would strengthen Locke.

From your summary, your breathless tribute
I am reminded of another, Tourte. [265]
François-Xavier spent his adolescence
With rod and bait, (peut-être fromage, baguette…)
Ambling alleyways about the Seine
Where fractured crates unloaded from the boats
Lay scattered in the early rising mists…
Homeward with a shoulder-load of splints
(Together with the lean or copious catch)
A fruitful day with raw material to make
The bows that were his father's modest trade.

Involution

Illiterate; no musician, he
Sampled timber from the Indies to Peru…
Perhaps too idle to ply blade to whittle curve,
He held them over charcoal and gradually
Persuaded them to bend across his knee…
Those that held an arc through heat
Proved lithe as snakes, resilient, biddable…
The sostenuto slide across the string would be
As tranquil as a billiard ball on felt.
Spiccato from the balance and the bounce…
A natural fulcrum at the centre, near the bridge
Tips out rivulets of notes that simply flow
Astride the gentle tempered call of gravitation…
This purist Stradivari of the bow
For Cremona was a marriage made in heaven.

A hidden modest dowel conveys from hair, through string,
The espressivo of the finger's fine judged weight;
The soundboard of the back, a figured maple,
Sends out music through the waiting aether
Connecting heart to music's heart but through the ear…
Vibration is the universal chain that travels
Through hair on gut, through wood, and taut membrane,
To tickle feet, suppress a sob, or prick a tear…
All from incidental debris on the Seine.

Something in the Zeitgeist shaped that course
When Pernambuco met the favoured Tourte.
(And Pau-Brazil gave its continent a name) [266]
Newton's laws of motion; friction, ricochet
Are music's Hermes spoken by a slow-grown wood.

So Newton tripped upon his comprehensive laws
From spillikins of Galileo and Descartes? [267]
(As Tourte refined his red and mobile stick)
With gravitation now the glowing coal?
Dissected out the components of white light
Through prism's scalpel spread the spectrum wide

Canto the Seventh: The Enlightenment and Rationality

To validate Descartes' intuitive sense that
Colour was not intrinsic, but derived?

As Tourte adapted his first stick of Pernambuco,
Gave its frog a brow to spread and flatten hair,
Devised a silver mounted screw to tighten tension,
A neck as shapely as a gliding swan…

So Newton laid the planks of bevelled law,
Fine sanded, mortised, dowelled to form a board,
Straight grained and running true as oak
On which three centuries of cooks would lay their dishes
(All applications of generic law):
Chromatography, the periodic table…
The elements within the planets, stars…
Doppler's whispered light, red shifted, spoke recession from
The origins of matter at the start.
Relationships of work and transmutation…
The chemistry of growth and respiration…
Closed circuitries of matter's conservation…
The cat's cradle of its language, mathematics
Squiggled sermons for the literate devout;
The new priesthood now would speak in riddles
But works and engines certainly impressed!

The Principals have, for the moment, left the stage—
The marriage between the mind and natural world
Is attended by a pilgrimage of crowd
Whose lines are improvised, 'ad libitum'
In a welter of whispers we discern
Factions that concur or disagree…
Universities across the continents' divide
(Paris, Padua, Cordoba, Leiden…)
Send guests to check on other's inspiration,
(Precious gifts are now a new idea
Often cipher wrapped, unreadable by thieves)

Involution

Newton has by mathematics proved
Man's mind is equal to his world;
Needs no instruction or authority
To investigate an issue prima facie.
Infant science now will stand erect
At liberty to wander or to trip,
Innocence remains clumsy and wide-eyed,
The polymath, amateur, men of leisure
Educated and with mostly private means
Fight for primacies of claim; take out patents…or
Meet in taverns in the company of friends.

We cannot hope to make collective sense
Until the fête champêtre has faded and dispersed.
This collective penetration and recall
Divides disciplines to quantify the world.
From the leavings, dishes that remain,
We may sample early ones to taste their flavour:
Solids are distinctively arrayed…
Small systems each in decorated china,
Seeming separate, crudités with crunch
Until melted in metabolism's butter
(In perpetual state of Heraclitus' flux…)
The rules are stretched, matter is conserved, but
Soon many cooks will serve us tepid broth.

The collective intellect is now in measures built
From divisions in the disciplines of thought;
Categories, hierarchies and tables,
(Karl Linnaeus to Dmitri Mendeleev…) [268]
Collate relationships of process or construction.
The museum of an ordered understanding
Is checked against the tolerant Creation…
Waiting for the intellect's submersion
In the drowning wide Sargasso sea of love. [269]

(Recall the gourd that metaphors the mind?
The diversity that stretched its swollen past?

Canto the Seventh: The Enlightenment and Rationality

The tree of life that spreading was mistaken
From man's perch upon time's incidental twig?
Now begins the collective recollection
(Newton has defined the common laws)
His successors now are left to paint the detail
In smaller intuitions that align
Relationships, forces, evolution,
(The closed history of our intricate machine)
Severing intellect from a fitting exaltation…
Instead a growth of greater separation
Through tools and progressive exploitation
Of a world at the mercy of man's mind)

Now who is jumping gun to premature conclusion?
Why not celebrate what you know best, the art?

I will in time, but first you have the floor to sample
Encrusted dishes on the smorgasbord of 'truth'

The appetizer to atomic theory [270]
(Corpuscles like mussels cooked by Boyle)
Whose law of volume proportional to pressure
Suggested 'something' suspended in the æther;
Irreducible 'elements' that were the 'universals'
Of manifested matter's diverse forms.

Epicurus and Democritus now caper
To the merry tune 'Did we not tell you so?'

Aggregates were by him called 'clusters',
Dalton would later dissect rules [271]
Along clean lines, shaved by Occam's razor, [272]
(The simplest option is usually followed first)
Molecules and compounds are combined…
Elements are linked in fixed proportions…
Affording substances of very steady habits
Families grouped as acids, bases, salts,
Inert, combustible or gaseous…

Involution

(Some dormant in blue bottles and small drawers)
Atomic number and the periodic table…
Completed after Mendeleev has searched,
(Just as gaps in orbits yield predicted planets…)
So the 'Skeptical Chymist' opened up structure
For atomic thinking that leads us to Niels Bohr. [273]

Between Boyle, Hooke and Locke
Britain ploughed up acres, [274]
(All bachelors, too wise to take to wife…)
All Royalists needing close discretion:
The 'Invisible College' met in secret chambers
Until it 'came out' under Royal Charter
To flaunt ideas down the catwalk of top fashion,
Met in Oxford to imbibe intriguing coffee,
Two knew Galileo, studied Greek, travelled wide.

Boyle writes the comprehensive 'Forms and Qualities'
(The foundation of all later chemistry)
Hooke, now called Britain's 'Leonardo'
Helps Wren on the structure of St Paul's…
Surveys London, destroyed by plague and fire…
Designs the Library of Pepys, the Hospice 'Bedlam'…
Anticipates a gradual evolution…
(Before Comte de Buffon, Goethe, Cuvier)
Is first to liken 'cells' to monk's enclosures…
Quite often (and bravely) takes on Newton,
(Light was not corpuscular but more like sound, a wave)
To relax he plays astronomer and measures
By parallax, the distance of a star…
Yet now perhaps is lastingly remembered
For fine drawings in his 'Micrographica'
Of the minutiae of such things as fleas and lice.

Locke extends Descartes and Newton;
Proposes liberty for habits of the State.
His 'Theory of Mind' gives simple observation
The power to oil the travel and direction,

Canto the Seventh: The Enlightenment and Rationality

Easy tolerance, trade, the rights of labour,
Medical ethics and the social contract…
The 'tabula rasa' of the free and spotless mind
Should be written only by the quill of reason,
Inked by dipping in the world of clearest sense.

Science was Philosophiæ Naturalis:
Voltaire, Montaigne and modest Locke,
Were widely read, sliced onion thought, still driven
Towards perfection and a better lasting 'good'.

Pertinent here perhaps to mention
The flow of thought from early origins—
A simple experiment would generate
Diversion for a dozen later fields…
These enthusiasts (like Greeks) were not constrained
By close counsels or myopias of interest: confidence
Took mental strolls at whim
Down byways or canals that looked attractive…
Perambulation later joins up mapping…
Terra Incognita's shrinking dragons
Would be tamed by a whetted mental sword.

The food of genius takes a slow digestion,
(Attendant guests will gradually disperse
To tongue the dissolving revelation
Of a world so integrated, whole and self-correcting,
That to spoon at its marrow or its essence
Seems to scar a new-born palpitating babe)
So beware of naming names, or listing people,
However meriting attention, they distract.
The wide sweeping broom of intellect's dissection
Is the subject of this treatise… consciousness
Takes any outstretched hand that reaches up.

The wave that gathers in a swelling line
Becomes translucent only at its breaking height;
The body of the ocean's ultramarine

Involution

Gathers glassy cobalt from the hinter sky
To break: In crescent urgencies of tumbling foam…
Sweeping greedy up low contours in swift surf…
The brine that soaks and disappears in sand
Has its origin from impulse far off-shore…
History impels the swelling wash
To fur in bladder wrack or beads of shell and lace
The coast of minds receptive and athirst.

So, stay detached: Observe the bunker table
Pinned with missives hot-footing from the front…
National pennants and insignia stripe-coded,
New platoons, spreading roots, joining hands
To pin creation's marvellous collection
In a growing quilt, appliquéd and close stitched
On a fine white-woven single linen ground.
The tabula rasa, unresisting,
Is covered by assertions, then corrections…
Counterpoint, complexity of fugue.…
We observe the great adventure,
The wave travelling, spreading is absorbed…
Evolution made its bumbling errors, likewise mind
Stumbles backwards through the story of its past.

(The narrow cone of earlier creation
Was simpler in its chaos, uniform:
The slow plunge of successive inspiration
Returns with plums of wider staining laws)

The vast warehouse of experimental thought
(From the molten planet's core
To the heat in distant star—
Both joined at time's conjectured start)
Stretches up an axis vertical,
Piercing through organic layered loam
That grows the single beanstalk that is man.

Canto the Seventh: The Enlightenment and Rationality

Along this axis looms set struts, establish claims,
A narrow warp, strip-limited in size, ²⁷⁵
(Set a-treddling by dogged single men);
Shuttles weave a closely argued weft
From alternates that batten right and left…
The fell proceeds for years quite undisturbed,
The hammock fabric seems sufficient, strong,
Until the selvage at the margins frays and breaks
For want of answers to new questions posited…
The warp is widened, disciplines connect
In deeper process and the need for common threads.

The quarry where geology began
Has now Stratigraphy to shake or plumb
The incremental layers, long laid down…
'Oceanic drift' say steady Neptunists,
'No, volcanic cataclysm' shouts the Vulcanist
(Whose holidays are spent on Krakatoa ²⁷⁶
Or Iceland's bubbling blisters hissing threats)
Palaeontology gathers up the crumbs,
Fashions skeletons from a calcium flake
Or a diet from a bowel turned to stone.
(Whole papers on an unexpected seed
Sends up flurries for the Paleobotanist…
And tectonic experts all to check their charts…)
Geomorphology lies in continental drift,
Geophysics takes the temperature of wind,
Makes adjustments for the oblate shape of earth:
Isotherms, magnets, meteors…
Von Humboldt deduced deep fissures spewing heat. ²⁷⁷
Mineralogy tablets thrown by Abraham Werner
Made him Moses at the Freiberg School of Mines.

Each took its tools, its carpet bag,
Devised, extended senses, as required.
Gay- Lussac needed altitude to study air ²⁷⁸
(Balloons saved legs from strenuous climbs…)
England's naval ambitions were much safer

Involution

With the Equatorial and Harrison's chronometer
For the man who went down to the seas again
(The lonely sea and the sky…)
The tall ship had a rudder constant feathered… for
Clockwork tracked the star to steer her by.   [279]

Each cell in the honeycomb of thought
Feeds, swells, fragments, and then dissolves.
Tools, which were initially assembled
(Cog of brass, a convex glass, or tilting plane)
Become the wavelengths of deconstructed matter,
Sonic, x-ray, infra-red, or laser
Date age of stars, the speed of their recession…
The elements of long antique existence…now
Slices bodies with a virtual blade
To find a cancer, cauterize infection
(But never seek its story or its cause)
We date our ancestry with everything that moves
By a bar code passed beneath a scanner…
Measure complex undivided systems
That cannot be reduced to salt or boiled
(Without destroying what it is we're after…)
Living systems shake a silent fist… or turn the other…
By giving up the smiling knowing ghost.

(Little black Sambo turned to melted ghee   [280]
Not the white-coat tigers all about to pounce…
Nor any pancakes offered for his trouble…
If only they had left him as he was.)

The human family once held racial labels,
Proportions of the features: bridge of nose,
The zygomatic arch, prognathous jaw,
Chemistry of sweat, the Asian eye,
How much melanin will stamp a person black,
The adipose extremes of the !Kung buttock…
Since eugenics was the favourite tune of Hitler,
Deprived the gene pool of the Twelve rich hybrid Tribes,

Canto the Seventh: The Enlightenment and Rationality

(And Einstein's brain no bigger than another)
Some kinds of thinking are no longer kosher,
Differences in men must be ignored.
Some knowledge is too dangerous to know.

Yet if science had conviction's courage, searched
In this, as in all other victim spheres;
Why the long shank Negro can run rings…
Why Orientals take to mathematics…
Why orchestras are filled with fiddling Jew…
Why Britain has snuffed its shining Watts…
Why France gives philosophers carte-blanche…
The study of the human race would focus and amalgamate
The common threads that generate
The universal longing to break bonds…
Escape the prison of the skin …
Barriers that block the streets…
The limits of habitual dreams…
Any separation that confines.

This universal hunger is ignored
What's so difficult about the urge to hurry home?

Vibration is really all there is,
Polarized in different ways…
Wavelength's spectra penetrate
Worlds beyond our narrow band.
Our humble cousins are au fait;
Bats could teach us orientation
(Through ultra-sonic high-speed travel),
Butterflies point out violet hues
In flowers we perceive as white.
Vision is a rough, not ready measure…
Most acute for narrow spectra…
Frequencies penetrate one another,
Spew, collide, polarity, colour…
(Neutrinos travel unimpeded
No charge, they rain like no tomorrow

Involution

Balance equations on both sides…
Confirm exactly what Pauli conceived
And quantum theory needed)

Light is our limit for a world of matter:
What then will measure speed or essence?
Or links in universal thought?

While you frame your answer let me take
A swift shimmy through concurrent art…
Your scientists are all at change and form
(The forces that play upon decay)
My painters look through lenses too—
The microscope of a searching heart.

Nature is 'objectively' perceived,
Studied, sketched, in studio or field.
No symbolic meaning given by the Church
(Or curtained for vain prurient wealth)
Nature is a turning world imbued,
Perfect in itself, all creatures too…
Such transport needs the self-effacing eye,
No message but its intricate deserts
Captured by the fluent brush or quill,
Framed, bound or played; to stand in simple shift
Reflecting back a worship that selects
The subject is himself 'Ich bin in dich,
Your essence is my own, and we are twin'.

The soul of art keeps the Platonic pace
Restoring wholes where science gives us parts…
It is beauty that it sets upon a plinth,
More intensely felt, more vivid, more complete.
Yet our split story of this journey is the same;
Man collective now explores his total world,
Its birth, its youth, its age, decrepitude,
Seasons, time, tides and changing skies,
The structure of man's features, muscles, bones,

Canto the Seventh: The Enlightenment and Rationality

His city, village, harbour, stubble row…

Art alone is conscious of his mind.

Painting sees uncertainty, despair,
Regret or innocence, disbelief or pride,
Companionship, the loneliness of deaf,
Poverty's burdens or its braced resolve,
Patience and the linen steady state
Of servitude.
These endow a human signature
To every subject, landscape or still life;
The running tide, the children, donkey, dog,
The artist strokes his canvas all with love.
Even the commissioned portrait bears
The painter's choice to show the features clear,
Redeeming those he cannot like with light;
His vision outlasts the sitter's life, or
Any memory or child he might bequeath.

The maid who fillets herrings for a meal
Or leans upon an evening windowsill
Is more immortal than presiding Pope
(On chair of state, behind a blinding cope)
Poised in her footfall on the path to death
Ennobled by the sunset's frozen clock.

The painter mediates artesian pressure
(The ebb and flow of human tides and wars)
Is filtered through his temperament and talent
(Tailored to the lining of his purse…)
The Renaissance had recovered Newton's laws:
The play of light, perspectives, depth of field…
Artists could make free with subject now
In composition, with but simple line and colour
To do justice the richness of their world.

Involution

Spain had religious sentiment in spades… [282]
Murillo offered liquid-eyed Madonnas…
Brave Goya was unswervingly satirical,
He skirmished with a palette he confined
To point guns at hypocrisies of clerics,
The unthinking violence of enlisted men;
Toledo sheltered the mystical el Greco
Lengthened by attenuated grief…
Figures all in imminent assumption,
Drawn upwards by celestial mechanics
From the harsh world now mired in the 'real'.

Velasquez shackled to the constipated Court
(Through pitiless studies of its routine habits)
Escaped its dressing rooms aswarm with dwarfs…
With drunken men on a Sunday feast,
Old women cooking eggs and chewing fat…
(His sharp reliefs would be grasped by Edward Manet
For direct engagement with emotional truth)
Later stretched by impressionism's capture
Of the essence (not appearance) on the canvas,
Until abstraction would melt boundaries and margins
Not merely of the picture but the talent,
The dog's breakfast chaos of a drunken Pollock, [283]
The circumspect reserve of Mondrian.

Dutch wealth, close buttoned, puritanical
Discouraged excess or overt devotion.
Painters migrated south like swallows…
Rubens was in Mantua well couched [284]
His painting free-handed, allegorical…
(Well, as free to labour as was Heracles)
His second wife was well upholstered…she
Exploded enough flesh to fill a canvas,
Politely now entitled 'Rubenesque'.

Van Dyke set the course of English painting [285]
At liberty with rich stipends from the king,

Canto the Seventh: The Enlightenment and Rationality

Whose vanity needed constant feeding
With horse and hound, pale and porcelain women,
(Hats of ostrich, modesties of lawn...)
Country life was the English aspic;
Parklands and the crumbling Gothic ruin
Satisfied its central contribution:
Painting could be mastered somewhere else.
(England snacked on plays and sonnets;
Pith of Dryden, vinegar of Pope...for proper food
Shakespeare was sufficient, quite enough)

Rembrandt stayed at home but never bowed— 286
He painted what he loved so ended poor:
Gentle Saskia as Flora or the Virgin,
Hendrickje wading in a simple shift,
Christ was restored as the Redeemer
For Amsterdam's impoverished disgraced.
His analytic eye kept tally on his features,
Mortality was measured by his face.
He modelled light with tender toleration
Of its pitiless erosion of bright youth...
The low country cannot claim his plaque or tower,
It lay beneath his stature at his feet...
His shadow spread beyond, into the future,
Immortal was the visionary von Rijn.

Desist. Enough.
Let me interrupt your blithe soliloquy
Wherein art rides a flowing milk white mare
With a trip-wire and a once respected name: David. 287
You claim no message underwrites or knocks
The artist in embrace with naked truth?
This regicide befriended Robespierre,
Laconic and indifferent to the wheel...
(Gentle Lavoisier bare-headed on the tumbrel)
Admitted by a hasty concierge in slippers
To draft the martyrdom of putrefying Marat
(In his bath like any Roman emperor

Involution

Caligula at tomorrow's gruesome list...)
Forced to abandon the 'Tennis Court and Oath'
(Half the characters had already met the Barber,
In the blood soaked basket all their heads had kissed)

Not content with glorifying terror,
(Or drinking with the poisonous Saint-Just)
He set about distorting Bonaparte
Whose Alpine crossing on a modest donkey
Was rendered on a snorting vaulting horse...
Low perspective made that stunted man a giant;
So much for your integrity of art! This cynical self-service
Led to Goebbels and the films of Riefenstahl. [288]

Your objection does my argument some service,
As you drew attention to the national casts to science...
So I have doffed my cap to weather and confinements
In scudding through the Netherlands and Spain.
Revolution's budding rose of expectations [289]
Lead to actions that seldom bear defence...
Where spontaneity in act took progress forward
Seldom does deliberation gain. The intellect
Is partial and self-serving. So art
In service to affairs of State or Edict
Is poster painted, hard edged, over simple;
Brutalism is the art of the dictator,
(Empty promises petrified in stone)

Jacques Louis David was painting out of period,
Cold images as tearless as a tomb.
His glory has the surface of bright lacquer...
When art accepts distortion of its impulse
It is mesmeric, coiled as muscled adder...
All language is Janus-faced and turns
To speaking both recession and advance.
So you oblige me to illustrate through music
(Not immune to distortions of this sort...
Marching bands have certainly done damage,

Canto the Seventh: The Enlightenment and Rationality

The single bugle is the usual coda...) But
The heart is not as readily persuaded
By the false, as is the intellect.

May I hang a new and glittering prize?
A Damocles to cleave an equipoise?
Our Waterloo of watchful ordnance ranked?
I'll call a midget trumpet to our joust...
Is Turner not a name to turn up light?   290
His palette drenched in light and naked else;
Steam, steel and galleons in its spray all melt...
A painter poet with a brush dissolved
In liquid light his citadels emerge,
Carthage or the Ship of Ulysses
Loom from the mists of London's grey
Evanescent vapours, sodden skies...
A waking dream? Or is it bloodbath day?
His guillotine, the sun, still deeper draws
Unblinking worlds into its drowning eye.

*

Put by your doubts and indignations,
Hitch up your skirts...I'll waist you in a whirl...hark
The bandstand Sabbath summons, come and dance.

If chaste mathematics is the classic Latin
Of intellect's manipulative skill...tis music
Speaks the simple history of soul,
The philosophy of the complex heart...
From rhythmic dances on a single fiddle,
The monotony on hide or clacking clog,
The wheeze of sackbut, shrill of pipe or tabor;
An instrument becomes the soul's expression
Rhythm national as a Scottish tartan,
Pedantic and absurd as Morris men...
(Others dance on swords or beat a clatter
They skip surrender flapping handkerchiefs)

Involution

The wildest gypsy has a Czardas revel,   **291**
Tangos for an Argentine assault,
Mazurkas or the slow alla Polacca,
The dreamy Dumka for the lost Ukraine.

All retained by classical composers,
(The natural dye that gives a rustic stripe)
The sarabande kept the court's decorum   **292**
As well as crinoline, fan or bustle-hoop.
The polonaise a whole world for Chopin,
The stately minuet whirled into a scherzo,
The dumka embraced by homesick Dvořák
In Iowa pining for his home.
Raw origins remain in new creations
As onion flavours almost any goulash,
Gumbo, daube, ragout or plain old Irish stew.

Bach, in his excursion into Köthen,   **293**
(Released from Weimar's solemn Sunday serge)
Wrapped dances in fresh bouquet suite collections,
Sprightly themes now wove them contrapuntal
In varied tempi, varied keys all linked…
Augmented harmonies and variation
The tune rode upon a team of willing steeds:
Flageolet, piccolo, oboe, natural trumpet,
The Percheron heavy horse, bassoon or bass…   **294**
(Mimicked by the Church's hybrid organ
In acrobatic manuals, stops and pedals)
Sonority was the calling card of Yahweh
Wild dances more in tune with fairs.

Haydn dressed the suite in sprigged quartet…   **295**
(Decorum was preserved for courtly manners)
Two fiddles intertwined a wealth of themes
Blended by the mediate viola,
The cello to give rhythm, pulse and ground
For movements that made deeper declaration
Bowed low, then solemn spoke the sarabande…

Canto the Seventh: The Enlightenment and Rationality

(Nosegay minuets and trios were preserved)

Mozart never 'grasped' subordination, ²⁹⁶
He gave all four voices equal chance to shine,
Threaded perfect balance tunes across the spectrum,
All punctuated, sang out loud and were sustained.
His first were dedicated to his erstwhile master
Whose gratitude was silence 'subito'
Never a joyful return to the 'da capo'
Haydn never wrote a new quartet.

For symphonic form Mozart was a Newton,
(At twelve he'd mastered orchestrating scores
Setting out proportion, rules and pattern,
Steady sequence in sonata form…)
Spilled notes on reams of parchment, uncorrected
(Coloraturas, with cleavage, worked their passage)
Opera was buffa or profound…
Symphonies scored inside a carriage
Were offered, by the travelling minstrel, wet…
String quartets were music served domestic…
Jewels on white napkins for his friends…
The tipping of the risqué bawdy basin
Hid his yearning for acceptance and respect,
Underneath his abundant spirited cheerfulness
He wept.

The Goliath that walked beneath a greatcoat ²⁹⁷
(Glowering as he fought to ride
His vehement and imperious nature
Down Voltaire's sunny new pollarded street),
In the forgotten forty eight he found his Euclid
(Bach was, by then, ignored and dead)
That laid bare all bones of skilful composition…
(Haydn taught him almost all the rest)
He poured over reams of fluent Mozart,
He mastered, was preparing to surpass…
Early symphonies enlarged, extended…

Involution

Musicians now bemused by frantic scherzos…
Tangled variations were unending…
(Marathons on Diabelli's theme)
Concertos embarked on heated conversations,
Held their own against the forte throng,
Soared to lift the singing individual
Above the weight of common cause.

Conventions all were stretched, elastic;
In the ninth he blended choirs
(Only humans could distil an Ode to Joy)
Tempestuous, grandiose or poignant,
Candid as the sun succeeding storm.
This Michelangelo of the aural spectrum
Put man and God in single exaltation
Worthy equals for the worship of his soul.

Complex music keeps the treading measure
Of science and hierarchic forms;
(Geniuses in both find one another
Only genius hears the future calling…)
Tonality was built in upward layers…
From the ground reverberating bass…
Rhythms syncopate, broken, silenced…
Agony held… in base-drum… punctured… pause…
The symphony strove to match creation
In its sweep, its vast dimensions,
Its rolling moods through time and space…
Its small voice… its turning enharmonic season…
Its Sonorous Authority.

Yet encroaching deafness steals:
Silence mocks and silence beckons,
The canvas on which all sound is written
Claims its deep priority.
The Prospero that scored Creation
Is alone, all revels ended,
The play concluded, ships have gone.

Canto the Seventh: The Enlightenment and Rationality

Ariel regained her heedless freedom...
No tricksy spirit sings or listens...
Only the wash of the twilight cries...

Stillness now the new persuasion
Imperious silence fills the ears...
The unmarked stave the clear horizon
Makes music's journey new, explicit;
To hollow the heart for a deeper longing,
To finger the strings of a naked thirst,
Prepare the gourd for joyful breaking...
Returning to silence its borrowed jewels of song.

His autograph on resignation,
His flourishing last biography
Murders carefully structured language,
Dismembers each and every limb,
Summons magical fragmentation,
Discordant harmonies dissolve...
The string is thrummed, hair sawn and broken...
'Sul ponticello' adds its cry...
Euphony is swift amputated...
Comfort whipped away by wind...
Oak melodies tear away in sections...
The raft of peril swamped by storms...
The soaring violin now scrabbles
For a blue unblemished sky...

Flares of tungsten pierce the darkness:
The afterimage, falling cadence
Draws the battered soul to lee.

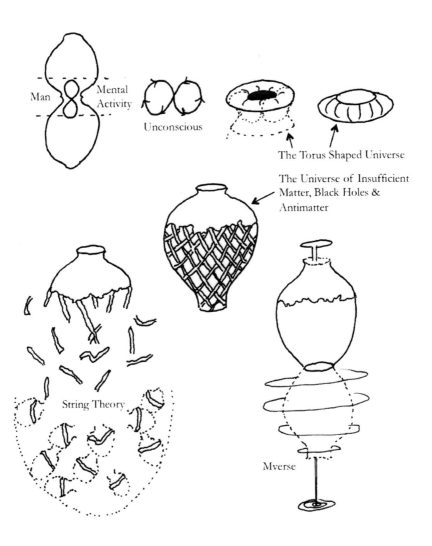

FIG.7 Voyages Round Infinity
Theories about the Universe
Structures and Shape

Canto the Eighth: Modernism and Dissolution

Your silence is most eloquent
Has music quelled your argument?

Argument was no intent:
The dialogue that we pursue
Is Socratic, nothing more. ²⁹⁸
Yet the nailing of persuasive fact
Is counterfeit, for sequencing instils a loss;
It gives to time false dominance.
Immediacy it quickly slaughters
By successions and chronology.
Each instant that propelled the thrust
Was instantaneous, a prompt.
The sun appears, and then it clouds…
Each genius that grasped a truth
Was both enlightened and a part
Of the swelling bellied mount…
Penetration now will narrow
To the apex at the start.

You took the wider wash of art,
Left me with logic, clipped and short.
Since now we must the world dissolve
Can we not sing in unison?

Stay but a moment. This last escarpment gives us pause,
Our dogged climb has much obscured.
This Kilimanjaro, one perfect cone ²⁹⁹
Is mind's creation; our ascent,
Our Grosse Fuge has been scored ³⁰⁰

Involution

In themes, that tracked, embellished, grew…
The deeper silence commandeered
Successive solos timely played
Just so. Like this. Another turn
Upon the wheel of twisting thought.

It is the monumental man that stands
Astride the cleft that separates…
One leg embedded in the clay
The other stepping clear beyond.
Act in the present alone creates,
Thought is the higher act we trace…
Take comfort, we have not far to go.

I am reminded of another: One [301]
Drawn to the pinnacle and offered all
If only He would kneel consent…
Shall I pine instead for innocence?
A lassitude invades; through all my limbs,
My head swims, of oxygen (or sense) deprived,
Starved by this altitude's thin gruel.
This peak is covered still with snow,
Its buttresses are steamed with cold,
Encircled with a meagre air…most punitive.
Sterile paths we must assault,
This landscape seems in permafrost.

Death of the mind is what you seek
This cone of intellect grows bleak.
But concentrate. Remind yourself
The purpose you embarked upon.

From those limitless sunlit plains below,
Those docile herds that ruminate
Methodically through golden grass…
(Unchanged as when we crept in stealth
Downwind with arrows, hammered flints)
We have raised up this terraced mount

Canto the Eighth: Modernism and Dissolution

By incremental excavation.
This is the slag pile of the mine
That delved to dig out memory
(Three million years since early man
Ten million more of ancestry)
Explosive blasts of inspiration
Cleared boulders in resistance clayed…

It seems solid now beneath our feet but wait—
The glacier above must melt.
It will dissolve the rock, the earth,
For we approach the narrow neck
That separates this perfect cone
From the mirror cone within.
This seeming-solid impervious world
Has a destiny it will protest;
The cataracts that soon will flow
Will tumble separation out,
All careful structures will succumb.

The proportions of the formal arts
Anticipate the boundaries smudged [302]
In successive knuckles bared
At rule, or orthodox assent.
Beethoven pulled at harmony
Stretched its chords to breaking point…
Schubert threaded poignant tunes
His longing caged within the bounds
Of classicism's disciplines…
Perhaps too searing to expose
Leant pensive on a changing key…
Always 'La belle dame sans merci' [303]

The high romantics, Brahms and Mahler, [304]
Wore hearts unashamedly on sleeves
Brandied melodies with plums…
Tchaikovsky tossed his Russian cloak
In sweeping thigh boots treading snow.

Involution

Brünnhilde was full frontal muse   305
For Wagner's loves and enmities,
No bounds at all, no quietude,
All raging human storms were staged.
Elgar walked the Malvern Hills   306
In very sober plus-four tweeds,
Close-gaitered in his English bonds
His sudden rapture bursts the cage…
Like wheeling flocks from lowering cloud.

The sated heart then quietly turned
To music spare and leaner boned:
Stravinsky, Prokofiev and Britten   307
Exposed the entrails of a score,
Their harmonies held cartilage,
Melodies were sharper edged…
The corpse of classical composition
Interred and gone. Schoenberg and Webern   308
The twelve tone scale, the bar-less stave
Dissolved all structures understood;
Now a concert lasts for days
Of a repeated, repeated, repeated phrase
Or John Cage composes silences.

Shall I continue?
Or am I done?

Might as well—
This was my tune but go ahead
I'd rather listen, makes a change…
This evidence could well persuade.

Painting needs no guided tour; you all know
Impressionisms fluid forms,   309
Pointillism's colour daubs
The eye interprets from the page.
The cubist's free abuse of depth,
Where back and front are single planed.

Canto the Eighth: Modernism and Dissolution

Abstraction foists its prior knowledge
On the eye, and roves unchained
Through all the schools called modernist.
The Renaissance has disappeared
In paintings that are framed ideas
Connected by the merest mist
To nature and its challenges;
Texture now sufficient image…
Or painted seeds to walk upon…   310
A bed in flagrant disarray…
(Encounter's sordid declaration)
Lighting switches off and on…
In empty rooms that no-one visits.

*

The Court has now the midnight shrug   311
Of its disappearing Jester…
Two spray cans man an I'm all gone
Images that take the mick
Rappers man now get in quick
I speak in verse or maybe song
Take em both and you ain't wrong
I can keep dis chat all day goin
Long as you listen man its flowin…
I get up de nose of all de natives
Dat's de price for bein creative…
We's waiting for the silly grey biddies
To bring out soup and sandwiches…
On the street man livin's high on de hog
For pertection you better get you a dog…
You got somethin better to say?
Speak out man? No? That's OK

*

The disconnect of mind from matter
(The elements now are fully stripped,

Involution

Frisked for any hidden treasure…
Nothing much put back together)
Do we better comprehend?
Now structure has been deconstructed?
Was the consequence preconceived
Of exile to this Mayday ball?
Lucifer's glittering promises,
The empty laughter cynical…
The garden through the open frames
Breathes wind through shifting draperies…
The darkened trees, the sailing moon,
The fountain's murmuring pleasantries…
The man and woman once embowered
With unicorns and nodding flowers
Take the air upon the floodlit terrace,
Gaze on Eden manicured:
Bury the stubs in an oleander
On returning, close the door.

*

We have traced the slow descent
From Serengeti's ruminants
Through Egypt, Athens, Alex, Rome…
Leviathan coils all sinuous
Terraced a mountain Dante-esque,
Searched out scribes to write, proclaim
What was already stored, and known.
Each descent (and up-thrust) did augment
This steepening slope, encoded coiled
In spirals tight in every cell…
Those ladders kept the onward tally,
Orchestrated each event,
Silently emitted absolute edict; Now
Un-coded comes a bugle blast.

Urgent news is leaked elsewhere!
Make haste for we must staunch its flow

Canto the Eighth: Modernism and Dissolution

Higgs boson is reported found [312]
(The mass-less God endowing mass)
The turtle back sustaining all…
(Perhaps the Aztec anaconda) [313]
It will prolong the false debate,
Plaster Plato's cave anew;
Shuttle thought creating corpuscles
Sprayed on the mist of infinity…
Go feed the fire, we'll reminisce…
The episodes that shadowed the play
To such a pass of emptiness.

I realize this curtails your plan
Or hastens it: To slowly trace
From this solid mountain of ascent
The dissolving of the intellect…
Reverse direction; take this event
(It was foreseen. Conception
Usually leads to birth)
So recreate its parentage,
(I have sterile forceps, sharpened sheers
To snip all excess, staunch your flow)
I will play midwife and assist.

Let us strip-the-willow dance, [314]
Weave our measured steps through those
That brought us to this very point.
The sub-atomic billiard balls
(Virtual trajectories making scratches)
Are dodgem-driven in collisions,
Their sparking signatures replaced,
Names are changed from snark to quark,
Numbers now are very strange
Spin is not rotation…so
The flow towards all thinking's source
Is all we trace as evidence.
Step the measure across the page
That we may better patterns show…

Involution

The slow motion plummet of the bird
That fished for food from side to side.
Broke the last ice though collective mind
From Theophrastus to Higgs (et al)
(Predictions now are made communal
Physics has its coalitions…
This Boson that swabs the deck of creation [315]
Will be called to the bridge of the Ship of Science
As soon as the sextant has been re-aligned)

*

>   Shall I commence?
>   And point in measures regular?
>   Poetry here will serve you ill
>   Didactic facts will quite suffice.

You can dress in showers of sparks
I will play the Deceiver's part [316]

>   *Theophrastus was with amber enamoured*
>   *(Elektron was its Grecian title)*
>   *The 'loadstone' had 'attracting' power.*

Petrus Peregrinus rather later
Showed that loadstone sunning its face
In the warm smile of the Polar Star.

Gilbert said the star was no lover,
The Earth itself was counter caller
A magnet in itself, with poles.

>   *Gilbert crossed the gap and claimed*
>   *Sulphur, crystal, resins, glass,*
>   *Had similar inbuilt tendencies.*

'Action at a distance' made it easier
For the collective now to swallow
Newton's general explanation
Of unknown nature, gravitation.

Canto the Eighth: Modernism and Dissolution

Guericke spun a sulphur ball;
It gave off sparks at each caress.
Hawkesbee connected up some copper
That shocked at any point he touched…
Current had to be in 'flow', and since
It moved equally to right and left
Two fluids had polarity.

Franklin spotted charge escaped
From wires and all insulation,
Sent needles quivering in expectation.
No fluids worked at one remove…so
Electricity was a miraculous 'force'
He drew it away like a magician
By metal rods on every tower
That drew a crackling God to earth.

Cavendish and Coulomb held the day…
Magnets and Electrics both obeyed
The inverse square law of attracting power,
(Provided charge is unalike.
Like charges, by the same, repel)
Remember: Opposition brought us here.

Galvani will help to nail this point
(By jumping to a wrong conclusion
From contractions in a pitiful leg)
That muscle generates a current…however
He was very nearly there.
It was the nerve that made it jump
Electrochemically.
Electricity is now fundamental.
The mind that shapes this comprehension
Is subject to the selfsame power
That is looking back.

Poisson lifted the spine of Newton's law
From its marriage to a wider world,
Then he showed its charms to both
Fields electrostatic and magnetic.
Mathematics now the feudal lord
That liked one wench as well as any.

Volta, an Italian, was more practical,
Leyden jar was now supplanted

Involution

> *By batteries, since called 'voltaic'*
> *(If experiments were going to continue*
> *Let science, at least, survive and print)*

Ampère was a spark that leapt
Igniting…reversed all previous views…
Magnetism was caused by current,
Loadstones held the spinning atoms
Aligning with the Planet's poles. [317]

Too large a shorting imagination;
His colleagues did not then catch fire,
Dense insulation is collective,
Time would pass to strip the wires.

Faraday, the dreamer, visualized…[318]
The Universe by forces patterned…
Iron filings (spread like waves) became
Invisible gradients of interaction.

Clark-Maxwell was a super-visualiser,
He dreamed in dynamic Technicolor.
All change in electric charge sent 'waves'
That travelled at light's speed through space.
'Fields' replaced the 'lines of force'.

'Electromagnetism' now united
Radiation all as gradients…
(From the visual spectrum of split white light
To ultra-violet, and infra-red
To ultra-short gamma and long radio).
Matter now became a tool
(The mirror reflecting its own structure)
Incubating a whole new infinite science…
(The invisible world of conceptual mind)
Particle physics will breed new gnats
Though filters like mosquito mesh
Will keep the little blighters out.

(Maxwell was content with mathematics
Abstraction was sufficient proof)

> *JJ Thompson now dug deeper* [319]
> *The electron was electricity's 'unit'*
> *And a primary particle of an atom.*

Canto the Eighth: Modernism and Dissolution

> *(How it escaped or did what it did*
> *Remained unexplored*
> *But made the connections*
> *Firing at so many cycles per second)*

To balance the negative charge now named,
Rutherford suggested a positive nucleus.

> The macro-universe of gravitation
> With planets in orbits round the sun;
> The micro-universe of shelled electrons
> Circling the nucleus with keep-away caution
> Were in Bohr's atom neatly bonded…
> (Mirror, mirror on the wall
> Miniatures and dinky toys
> Disappear like Russian dolls…)
> A solar system giving rise
> To everything; it satisfies.
>
> *(Images, images, visualization*
> *Since Pythagoras and Kepler*
> *Seeking perfection…*
> *What has changed?)*

> > *It did not last. Why would electrons*
> > *Never spiral inwards? What kept*
> > *Them going, and apart? Electrons had*
> > *A very small mass… by contrast*
> > *Its partner (the positive nucleus)*
> > *Was a sumo prize fighter and very large.*
> > *(Dynamics and structure were falling apart)*

Rutherford conjectured
(Invention is the mind's solution
When old ideas become a sieve)
Proton mass was not enough.
He proposed the existence of neutrons
Inert as to charge and therefore hidden.
Which Chadwick, as 'it,' obligingly found.

> > *Max Planck believed the orbits were 'shells'*
> > *Packets absorbed or emitted in 'quanta',*
> > *Electrons jumped from orbit to orbit,*
> > *What they gave off, or sometimes required*
> > *Were later called 'photons', the particle light.*

Involution

                    (One single photon is clearly visible)
                    Planck's constant, as 'h' now enters equations
                    $E=hv$ gave birth to the theory
                    That Einstein supported, but later hated,
                    'God plays no dice' is frequently quoted.

        Here we arrive at another synthesis-
        This will split all rational reason-
        Quantum Mechanics now takes root.
        De Broglie and Schrödinger talk of wave function... [320]
        Particles are not particulate...

Max Born objects. Waves [321]
Are not a particle's 'nature'
Merely the appearance of different 'states.'
Probability (the mind's uncertainty)
Will predict its likely 'atomic orbital'
Atoms will take different shapes.

                (Among these shapes he suggested a donut;
                The 'torus' is now the shape of the universe.
                At least it was when I last looked)
                The micro universe of the atom
                Now extends its conceptual reach
                Reversing Bohr's derivative atom
                To the macro one of a thousand suns)

                            Einstein was otherwise distracted [322]
                            (He did not care for quantum events)
                            Riding light within a vacuum
                            And watching all clocks slow...
                            And stop.
                            Space and time were better bonded
                            And space-time was undoubtedly curved...
                            The 'lazy dog' had suddenly awoken
                            To scratch at the itch of a narrow word.
                            Words meant exactly what he intended, [323]
                            (He claimed he was a slow beginner
                            Learning late his alphabet)
                            Relativity came from never quite knowing
                            On which side of a fence a crazy man sat.
                            ('It depends, it depends' cried the Professor
                            Swivelling towards and away from the light)
                            Humpty Dumpty fell off wall Newton
                            Its coping has never been mended since.

Canto the Eighth: Modernism and Dissolution

> *Heisenberg cools the physicist's clamour*
> *With a 'Principle of 'Uncertainty'...* 324
> *(There's profit in prevarication*
> *Have it both ways) Particle? Wave?*
> *Just decide before you investigate...*
> *Seeking position will produce a particle*
> *Wanting momentum will give you a wave.*
> *(Once given a title, it padded equations*
> *Curled up, and gave no further trouble*
> *The pin of the mind is working loose.)*

Relativity and Quantum Mechanics
Share a very uneasy bed
Both go daily about their business
But when they sleep it's back to back. 325

*

We are almost done with mind's a-drilling
From these oblivious herds to the source of time.
The closer man gets to the start of creation
The more he oscillates 'either-or'.
Since then a cloud of elementary particles,
Six flavours of quark, six kinds of lepton,
Twelve gauge bosons to account for the forces
Weak and Strong bonding the atom...
(New Apostles are given clear instructions
Remember those virgins, keep alert)
So we can invest in a Hadron Collider 326
To light the Bridegroom's sparkler path...
Just give us chance for new auditions
To test this queue of hypotheses...

But now we have another problem,
The violation of symmetries.
Nobody knows why there is so much matter...but
Of anti-matter not enough.
Now matter is the smile of the Cheshire
In a bubble burst hole in a vacuum tree.
Higgs' boson might provide the answer

Involution

Or else our model must be abandoned…
Conservation has been the be-all of theory
(Newton has served us very well
Whatever the natural world manufactures
The opposite must in equals exist…
The intellect has the self-same reflection
It was built, after all, by opposing itself)
Unless the answer lies in 'strings',
The invisible dimensions we cannot measure
Invented ad infinitum…
Imagination seems to keep us going
But matter simply disappears?

Do I detect a smile of pleasure?
Or hark to the sound of a rumbling purr?

The cone of intellect subsides,
The mountain melts and all is plain;
A sea of radiation covers
The matrix is alive with sparks.
Space-time is now the only dimension;
All is here, the past, the future,
Thought creates, but instantly.
Translation is no longer needed
Through the laggard world of matter
Slow to manifest, obstinate…

What are those intricate spirals twisting?
Those miniature Leviathans?

Those traces of the mountain's history?
Released from every melted cell?
Those are your 'strings'
All the dimensions to which you are blind
(Every other's enclosed mind?)

Each has pursued a different trajectory
As varied as the leaves on a tree,

Canto the Eighth: Modernism and Dissolution

*All combined to collude in illusion
We, hitherto, called 'Reality'
Asymmetry lies in the positive image
Of coloured nature, beyond and 'outside'
Without the negative offered by mind.
(The shutter of the intellect's camera
Developed only half the narrative
In the tank of corrosive certainty)*

*Defied at intervals by spontaneous instants
When mind met itself on the stair coming down…
Or sparked another in love or rapture…
Space-time's injections created creation
Across all bonds of language or nation,
Sailing the seas in the teeth of time.
(The Soul of the Universe is bonded by spirit
Merely a word that works in Einsteins)
Spirit lies buried without a headstone:
The negative matter sustaining creation,
The dark photons that light the firefly path,
Igniting the genuine searching heart,
That throws all the bolts, drops the drawbridge,
Opens the portals and lets in the light.*

*

*Our journey now is almost done:
We approach the river Jordan
Where our dialogue must cease.
Reason, friend, you may not come
Beyond the water's flowing course…
Our camels here we will release
To sink their pace in folding knees,
Unpack the travel's bundled load
Bury the hatchets in the sand.* [327]

*You have entertained, but now
I must cross this stream alone,*

Involution

*For I return from whence I came
(We were companions for a time)*

*We scrutinised creation's egg
Through the prism of man's mind;
The ocular bent narrative
Distorted, in parts, by your desire
To detail logic's path to here.
(You invited me to shine the torch
Upon selected heads in light
Science's inspired men,
Music's genius, painting's art)
As though the beacons they each lit
Would flash companions to this tent.
(As would a flare in desert night
Set a weaving staggered course
Towards encampments in distress
Starved of water, deep in drought)*

*It may indeed, but should it fail,
We have traced through oases
The subterranean welling force
Of memory's own summoning
From undivided consciousness…
(In the Eden of the Aegean
Where light most incandescent shone
Mind immortalized in temples
Language, poetry and stone)
Through dissection's grafted tree…
Convergence here is where we part.
No division here remains
Within the world of Nature's rich
Manifest embroideries…
Only man has been deprived
Of equal joyful harmonies.
For certainty has been secured
By severing that tree's deep artery.*

Canto the Eighth: Modernism and Dissolution

Your travail does eliminate
The plea that in mankind recurs,
'If only we had known or seen
How differently we might have walked!'
This epic, from its honest start
Declared its intention to repeat
The travelogues of other poets…
All perennial philosophy…
Just another candle lit.
In new constructed science sconce,
Iron guarded and well oiled
To taper light above those heads
That passed beneath and unobserved.

One tread is all it will support,
All crossings made in single file
And in thoughtful solitude.

*

Our life has been your Bible book,
Forsooth your bell and candle, both…   328
(For life is what this treatise probes)
Other pages, other lives, here's the library interred.
You read your memory's catacomb
As a fluent blind man's Braille…
Has darkening the lesser light
Dipped a quill in lemon ink?
A sudden match-flare revelation
To reveal a future's clarion?
There is, within this narrative, a holding of the breath…
A sense of promise spilling forth, a pregnancy of child…
Three gifts are here implicit: A certainty I never met,
An incubating joy, the patience giving liberty
Some time to comprehend.

It was all I had to bring
For words must pull against the grain

Involution

Shackled to their past
Employed in other houses, poorly trained…
Blacking scuttles, shining brass…
All marred like aprons with old stains
Or rigid in a tail and bow.
Unable to use a common tongue…

Upstairs downstairs, pass on the stair, aliens estranged:
Only Cook creates and mutters ('my secrets are my business, child')
The Gardener's peaty Saxon boots (that are discarded in the porch
He speaks to plants throughout the day, and has been loved enough)
So you spoke art and music's flow…
(Music talks in rapid Italian: Ballet in most proper French,
Art hardly ever speaks at all, but hopes to hear a sudden gasp)
But Science is the pompous butler
Replacing all the corks too soon…
Closing all the shutters…
Against the evening's benison, the twilight's long suspended light…
The children dispatched to nurseries (before they spill the saucer moon)
Leaving the Turkey carpet cluttered
With Aslan's wood, the Ice Queen's tower…
The Meccano litter of their game
He sets to rights in a slotted box.

Through our joint excursion (and with your help)
I hope that I have shown
That future it was that called to men
Attuned to its quiet voice…
'Knowledge is of yourself, within, and of all else besides…
You are the microcosm of it all. Submit and you will find'
The upwelling of our origins, early speculators washed
A broad palette of initial colours that all together blend…
The bubbling fountain overflowed…consciousness
Unfiltered through the mesh

Mind and matter were two sides
Of an ordered Universe.
Man stood astride the rivulet

Entrusted by a faith
That he would honour and vouchsafe
Care and husbandry to both.

Then the cleft fell like an axe!
Split Eden's innocence,
Sent man into exile with a brain
With which he could survive
By mastery of tools and eyes
That watched the world without.
(Emissaries came occasionally
To check his wayward path…)
Until he traced full circle (as any planet orbits sun),
Old soldier with his bandaged eyes
Returns limping to what he once called home
And does not recognize…

*

Yet secluded solitaries in cells
(Those mavericks continue) [329]
Find the cosmic harmonies renewed
In waving integration.
The hexagrammic water drops…
The death-toll conflagrations…
The civilizations that collapsed…
An institution's natural span…
The seismic cataclysm…
The holographic Universe [330]
Now winks in fractal pattern
Where each component holds the whole…
Small is beautiful once more, [331]
The gem reflects in facets.

Kepler's tunes are newly sung,
All coheres; perfection's
Mathematics of the mind
Was always intuition

Involution

That microcosm in the man (the macrocosm's fractal)
Held the secrets of the Universe
Heard in contemplation.

Memory has walked us back
To origins forgotten.
Pythagoras has come to dine
With Plato as companion.
Musicians tune his Universe
His octave soup in fractions
The fundamental harmonies (he heard in the Aegean)
Were suspended till maturity
Clamoured for the cadence.

(Enthusiastic adolescence
Took apart an engine
To find a ghost alive and well
Singing like the siren…
It sings, but not yet lullabies
For the nurturing is absent…)

If science does our tale dismiss
There are libraries besides:
The Masters of the glimmering East
Vedas and Upanishads… [332]
He of Assisi, who talked to birds,
Rumi's whirling stillness, all mystic narratives… [333]
Science now might doff its cap,
To collective inspiration
Of genius that knew answers but confessed
Not how they were achieved…
These adventurers before the mast
Who frequently went mad
For want of language to impart;
Certainty so deeply dyed nothing bleached it out….

Canto the Eighth: Modernism and Dissolution

Shamans with narcotics,
Anaesthetize the mind
So better through attentive ears
To hear the serpent sing.

I have hereby introduced myself,
From here you are dismissed;
You may remain and listen…but
From now please still your tongue.

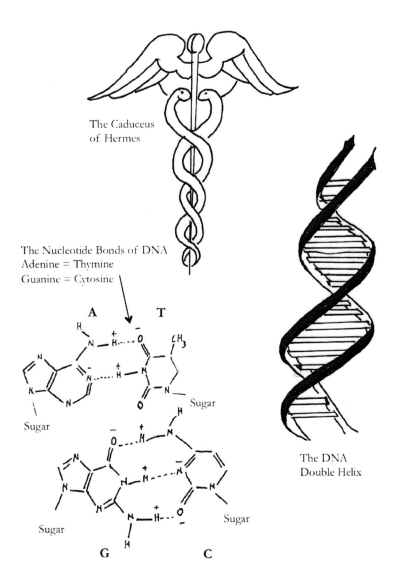

FIG.8. Variations on DNA

Canto the Ninth: Love and Reunion

I am a herald sent to call
You to your future home.
How may I entice you there?
Along this thread divide?
This fragile coil of argument,
Through muslin shroud of mind?

This soliloquy is my siren song
I must myself unclothe [334]
(Though Reason led you by the hand
I sing but to an ear)
I have in Reason's solid tread walked
In footfall's anchorage; but now
Caduceus is bid [335]
As Prospero's new enlivened staff
To conclude this narrative.

I would waft aromas sweet, a scented Amaryllis
Brushed with myrrh, tinged with rose
Just beneath your nose. The spice (or sweetness)
Would persuade your waiting feet to follow…
Have you tentative lick this page (of cream parchment fine as linen)
To taste ambrosia and with the tongue tip out the tang
From the cleft of your surprise.
Entice your curiosity by a pipe upon the air
(A Pan for you exclusively…)
Have you rejoice that rocks attend;
That trees do weep and birds remain
Transfixed in listening.

Involution

Above all I would have you feel
My arm about your shoulder,
So we might both like lovers walk
Towards our certain future, where only our companionship
Draws a synchronous unwinding…
Travelling not hopefully, because we have arrived
To wander in the bramble growth,
Through the medlar garden [336]
Where all is lit from inside light
And apples not forbidden.

Yet I am denied you; for you do not know me yet.
I can only beckon through your mind
But hope to find affection.
Reason I perforce dismissed:
Now we are alone.
I must speak across the synapse
Of a single brain.

It limits all my utterances,
Shrinks my palette (limitless)
To words it is accustomed to;
Images that are derived
From lives it has retained…
Seduce, across these narrow bands,
The one whose bands I cannot know
Since I have not heard you speak,
(Not in this temple anyway) You may prefer
A richer frangipani scent? [337]
Ambrosia not your food of choice?
Rather like a Crème Brulée, very nice, but everyday?
I can leave it yet un-named.
And offer rows of dots………

This mind is attentive through the ear
(The cradle rocks and swings)
The dialects of disciplines
Left concepts gathered on the wing

Canto the Ninth: Love and Reunion

To serve the pipe through which to sing
An intriguing narrative…
That echoes in the after air
Will remain, translated there…
So when this book is closed
Some essence of its soft address
Will stay; persuasive, comforting.

All workmen are by clay confined
By timber shaped by wind,
Or centuries of drought…
I can only manufacture thought
Through this given mind…
I would, through this interpreter,
Lead you only to yourself.

Reason has done her best, delivered you
Through history of stepping stones, to this estuary…
Let me paint the picture now
(The point at which we stand)
Climb up this hide that watchers use
For spotting geese and grebe;
Together let us, side by side,
Review the journey we have made.

Before us slips the water sheet,
A silk taffeta of spread,
A flow so deep, inexorable,
There are no smaller signs; no wave
No splash, no ripple sound…
It pulls as does a birthing child
Towards the light and air…
No distance now, the murmuring
Smells of salt and weed,
The sun contributes skin and heat…
The arms flung out
Of its embracing sea.

Involution

Look to our right, its silver shine,
Its clear expectancy…
The line of its horizon floats…
Does it not move against the eye?
Ride sight as would a buoy salute
A disappearing ship?
Unmoved by all disturbances;
A womb for all Earth's continents,
Her arms about all shores…
The sea, her salt, her shrugging waves,
Remains impervious.

See to our left the land traversed
Abased, for it is passed.
From here appearing estuarine
With sweeps of rattling sedge.
The outline trees against the sky are tonal, grey as smoke
In watercolour's rapid wash, suggestion is enough.

Across the river is the bar
(The gate of integration)
Where fresh water seeks submission
And salt is stronger in its claim
For avid dissolution.
Warm as Ethiopia, its rust light
Bathes all in equal bath; no details clear
No structures up, a honeyed spit of sand…
That is our destination, hold my hand
I shall conduct you there.
For it is all well known to me;
My cave is deep below…
Reason has cast the baited line
To anchor a tentative walkway, just
By stroking a pencil through time.

Two shallow curved parabolas
Are sprung suspension steels
(The narratives we traced—

Canto the Ninth: Love and Reunion

That journey which she rapid sketched
In struts of modest pine. Those opposites support the treads
Man collectively has crossed)
I needs must now put rivets down,
Or else this epic melts.

Behind us will come other minds
Recasting all in oak…
Filling its gaps with detailed plate
A stanchion bridge to stand.
I do enjoin you, just enjoy
The flowing river below…
The sway, the pendulum that swings…
You will not lose sight of land
Nor yet the constant sight of sea
Reminding you of Turner's brush, it seems to come and go…
All pigments it contains take turns
Pronouncing each its claim. The seamless sky above receives…
Released by dawn to the egret's beak—
What else will comfort you with me
But dome of sea and sky?

This fragile footbridge has been cast
By minds that stood athwart
Creation's mind (in everything) and mirror intellect.
The running ropes connected them, unerring, to each other
Across times and different continents;
Through standing on a trusted shoulder
Each acrobat saw further…[338]
The palaeontology of mind,
The strata of its seasons…
Their complexity, forms and hierarchies
Filed and separated.

The museum of the intellect
Has completed its collection
From at-onement with the universe
Through all appearance of division.

Involution

Now we see the vanishing point, where parallels converge:
Where e will equal mc squared as daily bread and butter.
One view will see a solid world,
For another it's transparent, [339]
The enlightened fall into the Void,
The seeker finds the Plenum…
Where mind absorbs the intellect (as the sea receives the river)
Consciousness will run throughout…
The field that moves through gradients
Of matter's interactions… each
A component of the painted cave contributing its colour.

This journey was to lead you here
Through the narrowing gorge…
Give me half a hand at least…for
I cannot compel your love.
For love it was, that built this bridge…
(Intellect has not noticed yet
For all the reasons, Reason gave—
Competition, opposition, the transfer to the cave)
Of all the jewels that caught the light,
Drew lovers to marvel and exult
Were locked in crania, out of sight
(We need to throw them open)

If we cross together, you and I,
I will introduce my burrow…
I am the serpent of the myths,
My shining scales I offered…
What bridge does it remind you of?
Poesy must flutter away
I now become skeletal.

The mathematics of my form, [340]
In symmetries of coil,
My alphabet of two hands placed [341]
In prayer against each other…
My linking spine of hydrogen

Canto the Ninth: Love and Reunion

*Is my asymmetry
(As is the spine and brain of man
That links us both together)
My pyramids are sometime set
To conically divide; to split
Before a seed is set,
To introduce another.*

My language speaks in trinities: 342
*The conversation I have held through time
(The prologue of the future)
Short injunctions are to matter sent
For protein manufacture…
The casing is no hollow gourd
Each cell repeats and echoes…
My deeper song is spider spread,*   343
*A web of consciousness;
Wonder is a struggling fly
Caught in resonance.*

*My cave is subterranean,
Its language is of molecules
Perfected minions.*   344
*They do my bidding flawlessly,
The absurdity of error!*

*Caduceus I once was called,
(It was a name I liked.
It wore the flash of connoisseur… fitting
Since I was your constant guide)
Serpent was pejorative, never to my taste…
But it has stuck. I'll live with that
Until you re-baptise.*

*Tempter, guide and secretary…
Mankind needed all my skill (rather more than others)
This Paradise is for the asking, follow me
I will show you patents: My records*

Involution

Are accurate, my pictures all original, ³⁴⁵
I give my jewelled scales to those
Who worship and submit.
I keep my secrets under guard…
Sometimes I pass thin whispers
Through apertures in robust walls
To beggars who politely ask
(For genius is messenger RNA ³⁴⁶
They come to take instruction)
Others who break in at night,
Or take excess of treasure,
Shout, 'Eureka' or 'I have found God'
They never are believed.

Man built his mirror Eden
From all the riches I endowed, but
Without consciousness, that river
That flows through my kingdom cave
Connecting all together… ³⁴⁷
Man's mind, as he politely calls it,
Is a rattling pastiche.

The anemone of the human brain
Combs through a sea of mind:
The skull as porous as a sponge
To soak ideas from passing tides
The syllables of words or shapes…
Or music scored on silent stave…
The needle placed upon the groove
Replays whenever brain is tuned
To Beethoven's vibration. ³⁴⁸

Its hemispheres are deep cross chased
To complement each other.
One seeks wide and scanning finds
The milk of truth and pattern.
The professor in the other half
Finds logic threads connected…

Canto the Ninth: Love and Reunion

Inspiration is from shorting points
(The silver pulsing nerves that run—
The serving maids to senses,
Using all the quickest routes
That have become habitual)
The occiput umbilicus (the reptilian stem)
Threads the needle with the snaking spine;
The oldest sage traverses me
To the 'higher' region. [349]

The cortex that explosive folds [350]
(The pastry case of reason)
The sea urchin with its million pins
Rotating through mind's medium
Takes signals from the atmosphere
(Including cells below my cave)
Interprets them in metaphors,
Conveys them comprehended.

All language in some way obscures,
(Hides light under a bushel)
Poets swat images passing blurred
To tease food out from syllable…
Through rhythms in their sequencing…
Some benign chance echoes… [351]
Seeking music's patronage
For history's encounters…

The language may be mathematics [352]
Or music's tonal colours,
A Madonna flecked in a meadow of grass, [353]
The Light of the World at the outer gate or
Sparking comets meteor…
The winking ladybird of a thought
Landing on a flower.
As I remain in cells confined,
Broadcasting to the Nation,
I am the crux, interpreter,

Involution

The keeper of the kissing gate
Where mind encounters wider mind
And mind sits down with matter.

Apart from disembodied mind
(With infinite dimensions)
I keep vigil at the other end:
Assign to egg an embryo…
Float lives upon the choppy waves
Of naive parentage.
Stable craft will steam away…
Others leak or list,
Find ports bracing or unwelcoming
They navigate or they succumb;
All shape the soul returning.
(The kiln that glazes a perfect vase
Cracks out all imperfection…)
The clay re-thrown on a thousand wheels
Refined through soul's discerning eyes
Preparing its ascension.

The hybrid cross so advertised
(As essential for the gene pool)
Lies much deeper than the egg and sperm
Gymnasia, physical or mental.
Average considered good enough
(Yet average is the ballast
Denying inspiration flight
Weighting the refusal)
Genius is often poor,
Or mentally unstable,
Weak in limb and sometimes starved,
Orphaned or neglected.
Intrepid in its path through life
Eschewing easy answers,
Walking always solitary:
By such has man ascended.

Canto the Ninth: Love and Reunion

*It is perhaps impolite to say
That all men are not equal.
Evolving towards higher forms
Continues unabated.
Some have hardly reached the mind
(Life is appetitive)
Some impairment hastens growth…
Short lives quick recycle…
Discrimination will select
The cut of cloth to promenade,
Poverty, wealth, parri passu
Towards the universal.*

*One man's fortune invites another
To envy or compassion.
All are each other's battlefields—
To gauntlet out the smaller self
By challenges compelling.
Starvation limits lessons taught
As does satiation.
Hunger is perhaps refined
(By satisfying lower claims)
For the wild hunger of the mind
To lose itself in ecstasy,
Surrender to the river;
The Apostles of the arts attempt
To draw men into the narrow straits
Of Lorelei's seduction.* ³⁵⁵

*(The interfering should have a care…
(Interference split the atom)
With stem cells, or any spliced repair:
In supplanting my authority,
Or by calling weakness, 'error')*

*Yet, since I have wisdom in some measure
(If I have not, who does?)
I know that nothing is for nothing*

Involution

(I am constantly renewed)
I have a universe of spies; ³⁵⁶
My colleagues in the atmosphere,
In the depths of ocean,
Emitting codes for everything
At the speed of thought.
They amend my records constantly,
What is fed in, is then returned,
Creation is extended.

Come, come, no need to look like that—
Darwin's run a marathon:
The old man's looking very tired
His pit-stops all are shuttered…
(His Beagle breasted many clues, ³⁵⁷
He annotated fathoms; plumbed
New depths of deeper time
Connecting them together)
He just missed half the process…his worshippers
Still kneel. I suppose I should be flattered
That they imagine I control—
(Imprisoned in a bunker)
This accelerating brisker speed
Through mindless trial and error.
Working from old patents…
As though the universe is buffered
For want of any steam or coal
Or leaping new connections.

All vassals penetrate my walls…
(We work collectively) ³⁵⁸
What is thought is everywhere,
What changes, changes all.
(Schrödinger's cat is alive and dead: ³⁵⁹
Imagine both and neither)
The possible is integral: past and future meet in me.
(The more complex consciousness becomes
The quicker is the rate of change

Canto the Ninth: Love and Reunion

The less mathematics matches)
The intricate discriminates,
Intention is more focused,
Learning is elastic,
Adaptation quickens…

The earlier an organ formed
The more difficult repair,
Perched upon its narrow stool,
Blimpish civil servant…
Performing as it always has, deaf to any plea.
What they call the 'reptilian brain'
Keeps respiration going…   360
The heart must beat…
The lung will lift…
All automata with a will
Only yogis conquer.
The sympathetic nervous centre…well
I suppose I should be pleased
Life depends on me.
So man imagines that his mind
Can be called his own.

Where was I? Ah yes…So

I annotate experience
Through each and every life man lives;
I played Touchstone long before
Great apes stood upon their legs
Or slept in spreading trees.
My potential, infinite at first
(For I had not found focus)
Has not been very much reduced
By matter's small exertions.
Mind takes more attention:
Initially slow going (tools then were incidental)
Things speeded when it strapped on spurs,
But still I led from Athens

Involution

Through human wars and arguments…
Offered water to his parching thirst…

Where once I offered flagons…
Now, demure, he sips from straws…
For my cave of darkness narrows to my own inconscience ³⁶¹
In identical bacteria, or viruses content with simple cell division.
(I waited for precision…we were all infants, after all)
When I was granted my own cell, I became contemplative…
I could control who came and went, in 'niggardly invention'. ³⁶².

We have already outlined how
My consciousness permitted acts
(Mechanical at first)
That fed cleverness in increments
(You can be sure I stored)
Such Hectors of intrepidness
Forged, so others would augment;
The mind of man no different
(Genius has taken him to both our origins)
We stand together once again
Awaiting Delphi's intercession. ³⁶³

Man has built his barren Eden
In dry stone intellect.
Reached my bright moat surrounding
(The thick membranes of his brain) ³⁶⁴
Charon and I often pass the time
Waiting for some custom
To ferry them across this stream,
(Some liken it to Styx)
Appropriate analogy,
For death's obligatory;
The death of separation,
One from himself, and one from all,
Melting is required.

Canto the Ninth: Love and Reunion

Perhaps some will be undaunted,
Breast my challenge, come and visit…
My door is narrow and takes courage
Shall I identify its guards?
My twin snakes, so intertwined, nothing separates…
Ever since they were wrested
(The King Cobras from the Pharaohs' crowns)
By Aesculapius…
They have been in demand…as Ouroboros, Quetzalcoatl, ³⁶⁵
Sito, Ronin, Yin and Yang.
Liana vines imitate; the ladder through the trees…
Ladders climb, that is their purpose,
Mine's eternally extended
By mantras, repetitious…yet
Each coil gives context, relevant:
Until now scant consulted.

Do my mirror chromosomes
'Superfluous' in every cell
Not take you any further?
Given a universe of mind
'In forms' a universe of matter?
Yet the mind is looking everywhere
(Below its heel, under its chair)
For its partner, anti-matter?
Yet here it is, a perfect string,
Matching matter in everything,
Concealed in every cell that grows
From every seed's reunion.

It records all human conduct ³⁶⁶
(That includes all thought)
How else would children master faster
Skills that once took centuries?
Unless that library's updated
In codes that newly feed?
Paganini, virtuoso, one 'enfant terrible'
Is now produced in quantities by the Japanese.

Involution

*The memories of conscious lives
All make their declarations…
(Body language subtle)
Physiognomy reveals…
Speaks one cell to every other…
'Trust this smile, that one is forced,
That upper lip is obstinate,
Those eyes too close together,
That throat is strangled with a lie,
That laughter innocent…'
I resonate from cell to cell;
In the gut I feel the sinking,
In the nape I raise the hair…
Long before the brain will weigh
And persuade it to ignore.*

*Amnesia is my portal guard:
Once the individual has rehearsed
His route through evolution—
The earlier it was laid down
(The bone buttons on the nanny's coat…
That shining pair of bright red shoes…
Long summers spent in forks of trees…)
Incandescent and quick frozen.
The tedium of middle life
Is usually compacted—
(The routines of dying hopes in queues…
The thrumming fingers keeping time…
With children all departed)
Then yesterday already blurred
Before tomorrow's breakfast.*

*The old devil takes it as it comes
(Not many enter willingly)
He relieves all visitors
Of needless old valises.
The panic that precedes a death
Slams a studded bastion…*

Canto the Ninth: Love and Reunion

Between your eden of the intellect
And my Eden: Consciousness.

To resume my invitation…yes
A kind of death's involved
To embrace me as your Lover
(I have been gentle all along)
Past visits have been ridiculed
By sceptics or the jealous…
You will abandon apprehension
Safety (fear's persuasive twin)
Adherence to a doctrine…
Or language, incense, ritual…
Come, if you will come at all,
Unsuspicious as a child. ³⁶⁷

We have brought you to my portal
Alone, solitude invades…
For I am you, and no other;
No other knows you as you are.
I alone hold your essence
(And the essence of all others)
All are welcome, all invited,
I was entrusted with all records
(Creation's library)
My current name is unromantic,
The modern Hermes…DNA
(My moniker—by a lack defined…regretfully ³⁶⁸
Romance has all but disappeared)

Gabriel I much preferred
When to the Virgin I brought child.
Needed only a single lily
To talcum a manger's straw…
(One transcript of our ladder ³⁶⁹
Is all that birth requires…)
Surrender alone requested
For divine impregnation.

Involution

Women carry that surrender
In willingness to bear a child:
That rose sends for re-encounter
Signals clear as pheromones   *370*
By which a butterfly finds another
Across a space of miles.
Women practise death in every birth,
Capitulating naturally…
For children are the kernels cracking
Both body and the yielding heart
From Creation's stem to stern.

Some men have been lifted
Into unsought ecstasy—
(Beatrice glimpsed, shone through a poem;
For Petrarch, Laura's bel pié
Wrote a song for everyday…)
All the Romances of the Rose,
Mother Nature: Mother Russia:
Pregnant Venus carved
By early man with bulbous belly…
To the universal Virgin (consoling, ever weeping)
Promising each and every man
The intimate rewards
Of his Soul upon her path.

The water in this amphora
You begin to doubt?
You find it lacks the swoon of wine?
Try listening with your heart.
Although Reason has departed
I ask you to recall
The central search for answers
Pattern in all number…
Harmony's proportions…
Orbits in the heavens…
Was threaded on perfection.
For something worthy of his worship

Canto the Ninth: Love and Reunion

(Man's ever un-slaked thirst)
Guides all creative thought.

Fibonacci's series laid out all succession [372]
That followed naught and one.
Initially an abstract game, the juggled calculi…
(Each the sum of the two previous…
Are you still surprised?)
Is now confirmed as Nature's walk
Throughout the natural world—
In forks in trees, leaves on stems,
Seeds setting on a flower head,
Pineapples, artichokes and dynasties
Of uncoiling ferns and cones.

Is mine is the golden spiral?
My perfect angle never aligns [373]
One above another…
Each rotation presents new faces
To the sun, by leaves or mind.
My divisions are first conical…
My cell is polarized…
As an early calling signal
For every waiting child.

I have triangles for Pythagoras, [374]
Their hypotenuses march
Along Fibonacci's sequences
With each alternate step.
My ur-alphabet, 'ur-elements'
(Earth, air, water, fire)
Their opposition is inbuilt
By structure and specific bonds…
Has mathematics always read
In the blue-print of my chemistry
The structured cosmos of all mind? [375]

Involution

My structure writ large in nature's pattern? 376
Small wonder mathematics knocks
To hear unending echo…
Economy and elegance are
The aesthete mathematician's
Guide to the True North.
The small breath that lies between us,
(His search for symmetry)
All evasions of my influence…
(He calls my voice Infinity,
Writes my hourglass all widdershins…
And goes to cool his temper)
It taps upon his shoulder
While he seeks for new distractions…
(Higgs boson or a realm of strings)
A Sufi would whirl him to the door
Where mathematics cease.

You find my claims presumptuous?
That comes from your reduction
In esteem for anything you fear to understand.
(Remember Newton: what attracts,
Repels with equal force—
Wary intellect is cleft in two…
One creates its ricochet: only love submits)
I will be cast out for blasphemy
(The divine, ipso facto, must remain a mystery)
What the intellect can understand
Becomes unworthy of the heart.

I am the recent country cousin
Of science's strict family,
Still with dirt beneath my nails,
A rustic too uncertain; a grammar
As yet unrefined…unwelcome
To sup with 'quality'.
Allocated the scullery,
(My miniscule dominion)

Canto the Ninth: Love and Reunion

Where the interest on my five percent
Makes orangutan or human.
Remaining coils along my tail
(For which it sees no purpose)
Is 'junk'; the relegated legacy
Of 'errors', since corrected.

Yet kinship writes that calling card ³⁷⁷
On what they have discarded:
The siblings in a family?
The nephew or the adoptee?
In the barcodes of my sinuous coils
The access through my swinging gate
Admits each soul to shape a temple.

Soul is a word not much used now,
(By theology so complicate: by sentiment disfigured,
A lucent floating jellyfish, with no means of propulsion)
Yet it hangs behind the mind, impromptu judge and juror…
Alive to every nuanced thought: reading cant, hypocrisy,
Evasion or temptation. More than conscience, it pursues
Honour, truth and balance. It has a truffle nose for good,
A shrinking from the evil. Sometimes it limps under the lash
Of appetite's dominion. Yet it keeps score and will repay
In even coin, an eye for eye.
It plays both scale and baton.

Nor has mind much made its presence felt ³⁷⁸
To intellect, that carving knife…
Selecting only boneless breast ³⁷⁹
From the crisply basted carcass
Whose life, no longer evident
Can therefore be neglected. But
If soul determines nesting site
In family, tribe or nation
Mankind is brought under the lamp
Of science's distinctions.

Involution

If I am the waxen plate,
A palimpsest of lives…
Impressed by narratives I'm told
To match the soul with parentage—
The hybrid of arriving past
I assign to future—
My homespun stripes speak dialects,
Kinship written on calling cards
Each according to their scripture…
The child is father to the man, each ensures
The safeguards to their hungers…
The correction of residual crimes…
The denial of appetites outgrown…
The shaping of their talents
Offers incense to the brazier burning
On the altar of mankind.

Each soul is one immortal whole
(Its energy vibration)
Particulate in its liberty
To choose what has been chosen:
The dynasties within the arts,
The families treading Shakespeare's boards,
Cremona's lines of luthiers,
The homing pigeons returning home
To exhaust their passions…
(Ardour is not infectious
Nor art sufficiently paid
To fake a false conviction.
The soul, passionate intrinsically,
Burns steady and sustained)

Precocious early limber child
Seeks guided incarnation;
Leopold Mozart, so reviled,
(As Commendatore immortalized)
Without both his virtues and his vices
(Esteeming the gift but shaving its glitter)

Canto the Ninth: Love and Reunion

Would Amadeus, born instead to a putz-mädchen,[380]
Have survived? Or offered man a note?
Or too many, in desperation?

Consciousness co-ordinates; (the universal constant)
As translator I interpret:
As secretary keep the traces…
At the gate I look both ways
To draw past towards the future…

Matters' causes and inflexions
Are like lychees peeled—
Integument of prickled skin,
The succulence of oyster,
Stone shiny as a new recruit
To resist digestive juices;
All understood, each analysed,
No questions left to answer.

My causes from the other side
Are as precise and beautiful…
Each creature perfect in its way
Though not all in all things able.
The earthworm does not regret its toil
Nor bees their industry,
The mayfly has one halcyon day,
Why should each man be equal?

My world differs in dynamic,
Like draws to like, instead.
(Matter works by opposition,
Intellect, so patterned, follows suit…
Primping in the mirror… [381]
With slothful time forever in arrears
Contriving petulant excuses while
Trying on its older sister's clothes)

Involution

Thus have I eluded recognition
(Synchrony dismissed as 'chance')
No polarities reveal my quietude
No tension gives off crackle…
Love blends without suspension
And lovers are absorbed:
Gilding new sung liturgies
Refreshing all the others down the years…

For soul is always single
Its language understood.
The moment man forgets himself
He finds his colloquy prepared
Already by the travelled.
To secure the precious rapture's spoils…
Circumstances shepherd…
(The books fall open at the page…
Messengers arrive with scrolls…
Maestro calls to hear the pupil)
In our timely penetration,
Space-time accepts the baton
At liberty to interrupt
The orchestra's cacophony
With sudden soaring tunes.

The commandments of the future
Are already written.
(Man much larger than he knows…)
When actions turn towards the light
Of balanced integration
He will hear the harmony sent
From all the choirs in creation…
(But I protest at interference
My wisdom confiscate;
They interrupt my dawn 'reveille' [382]
By precipitous 'last post')

Canto the Ninth: Love and Reunion

(One life is but a siding
Where hunger visits barrow boys
Proffering charms or education:
Little for nomads on the move…
The poor all beg with outstretched hands…
For callipers to straighten legs…
Crutches to swing the amputee…
Others swift cycle through the throng…
Each paced by my instructions.
One death is the Master's signal raised—
The rattle and the final steam,
The white shroud over the empty rails,
The tail-gate blinking round the curve,
The train is moving on…)

Yet I am far more complex, mysterious and subtle… ³⁸³
Yes, I hold all of history, but all the future too:
(My language universal)
I rain down light, and by light reign
In realms unvisited.
Do not dismiss too easily my God like qualities.
Immanent in each and all,
(I exalt the individual)
Transcendent through the whole, ³⁸⁴
*Omnipotent, Eternal,*³⁸⁵
Have I said enough?
Yet I lack one critical dimension: Reciprocated love.

*

When you are ready, I shall pluck you
From brittle anonymity…
We'll meet where Petrarch first saw Laura
Abelard uncovered Héloïse. ³⁸⁶
It is your home and only I shall enter,
All corridors and shadows I know well.
Did I not stand behind the pallid lover tracing
The blue transparent veins along your arm?

Involution

Was I not present when you recognized the absence
Of worship in his strategies of love?

Remember how you felt consoling fingers
Through your hair when you turned to me in pain?
When morning bright you gathered up decorum
Did I not stand behind your straightened chair?
How often did I fill the glass you scarcely noticed?
While you were thinking of another love you knew?
How many were usurpers, and mistaken?
For each, in turn, I filled your mouth with song…
For each prepared the bed with smoother linen…
The drawn thread work of despair and disappointment
Made lace (and deeper longing) sharper creased.

Did I not walk beside you when, in anger,
You marched the stony road towards the hill?
The cloud and mists came cloaking round to hide you…
Recall the piercing sun that struck the water?
The medallion silver lake lay in your palm…
You were flooded with delight to be alone.

I know everything there is to know about you,
I thread your necklaces from all the beads of dream:
I feel your surreptitious itch to dance the Charleston,
I know the jingle that causes you to frown.
I am the rock that shapes your private dwelling,
The path through thrift you tread in barefoot sand…
I have placed my gifts daily on your doorstep…
The doe I sent to nibble at your roses?
The flight of starlings to salute your crown?

I have waited aeons for your recognition.
My perfect world requires your completion…
In your tossing bed I lie beside you
To watch in sleep your quiet breathing fall…

*

Canto the Ninth: Love and Reunion

I shall know the moment I may turn and lift you…
My hands will liquid shape your acquiescence:
In the silent break of day, upon my shoulder
Upon dawn's clavicle your happy cheek will lean,
Cradled in my neck you'll breathe our essence:
I shall carry you entwined and carefully
Through the silver light and striding water…
Wade until we drown in salt bright sea.

*

The Lord of the Dance you will remember
(If you return from our sweet ecstasy)
Betimes I will in undergrowth be hidden—
A full aloneness now will be companion,
Fearlessness will take your other arm…
For knowledge you will have the perfect measure:
You find mothers, sisters, daughters in all women,
Fathers and a friend in Everyman.
You weep only at the multitude's indifference,
Its blindness to the glory that I am.

Appendix

Mysticism and Science

Out of necessity, the conventional length of a book, we have taken a gallop through time, through disciplines and departments of knowledge, all the while failing to put on a serious face. Orthodox science is very straight-laced; perhaps because it is aware it has painted itself into a corner. What I hoped to demonstrate was the process by which mankind became more certain of less and less, and why there will seem to be no answers unless science acknowledges its blindness to its own inspired origins, to its chronology of recovery, and to its patterns of examination. I attempted, through the history of science, to acknowledge science's inspirations, but also to demonstrate its limitations, and to indicate that science itself has constructed the closed door before which it now stands. My hope was to offer new keys, instead of the old patents that it pointlessly continues to force.

Unlike Darwin, who collected specimens and observed their likenesses and differences before he found the explanation that satisfied him, I have collected not differences but constancies, very verbal origins, and palpable consequences. For the moment, I begin by quoting William James, whose comprehensive book *The Varieties of Religious Experience: A Study in Human Nature* ends a chapter on mysticism with this observation:

> It must always remain an open question whether mystical states may not possibly be such superior points of view, windows through which the mind looks out upon a more extensive and inclusive world. ... Mystical states indeed wield no authority due simply to their being mystical states, but the higher ones point in directions to which the religious sentiments of non-

mystical men incline. They tell of the supremacy of the ideal, of vastness, of union, of safety and of rest. ... The supernaturalism and optimism to which they would persuade us may be after all the truest of insights into the meaning of this life.

Saints and Scientists
'Oh, the little more, and how much it is!
And the little less, and what worlds away!'
—Robert Browning, *By the Fire-side*

The 'little more' in science has taken some panning out. It has given us most of our poets, many of our painters and musicians, and the pebbles in their holy streams needed no hammers to expose the gold of visions. In the gravel beds of science, much shaking is needed, for visions are partial and admixed, as is this account, with narrow necessity.

The knots in the rope we retraced through the cave were the names of famous men, limited for the most part, to the arena of science. By the dawn of the twentieth century, science had mechanized the materialistic universe, but since Einstein all has dissolved. We now are faced with dark matter and dark energy, the nature of which we do not understand, nor whether it is constant or in a process of continuous creation. The separate existence of the matter we do think we understand must now be questioned. The slow melting of matter has collectively approached those silent others: mystics who have long reported the dissolution of intellect, and who, throughout history, experienced submersion in the sea of infinite consciousness. Their experience of total being is also the experience of the non-being of the material world. In infinity, duality ends, reason evaporates, and the tongue is stilled.

I ended the cantos with the lone serpent, DNA, inviting the solitary visitor. This may be poetic license or may allude to the gateway where morphic fields interact with forms, but its metaphorical importance is deeper and more important. The mystic's path has always been a solitary pursuit. Sometimes that pursuit is rewarded by an experience that is inexpressible, for in it there is no 'one' to observe, and no stance from which to take a reading. Like salt in the sea, the intellect dissolves. Afterwards, there is only the evanescent memory of the dissolution of separation. Yet it renders the individual more aware of his precious uniqueness, even though he joins others of the same stripe. Regarded from outside, generalities always flatten and dissipate. From the inner standpoint of DNA, it is the individual within the context of all—of other forms, of time past, and of future summons—that is twisted together, valued, and expressed. Not the five percent that dictates protein manufacture, but the remainder, recording the past and projecting the future of consciousness, and visiting the brain on occasion in dreams, visions, and contemplation.

Mystical experience threads through all history, all countries, and all religions. Indeed, religious institutions are built on the deep foundations of an ineffable truth, and the certain dedication it always evoked. In *Mysticism*, Evelyn Underhill suggests the great eras of mystical invigoration always followed the great intellectual and artistically creative eras:

> When science, politics, literature and the arts, the domination of nature and the ordering of life have risen to their height, the mystic comes to the front, snatches the torch and carries it on. It is almost as if he were humanity's finest flower, the product at which each great creative period of the race had aimed.

When intellectual achievement seems exhausted, recurrent invigoration is recaptured by the individual, whether mystic or contemplative scientist. Now, the individual crying in the wilderness will no longer be enough, the echoes are too repetitive to remain unheeded, the dangers of scientific blindness have become too pressing. I believe science has exhausted its first era, built its history from memory but at the price of its own, and the Earth's, near destruction. Reappraisal is now urgently needed.

William James's exhaustive study *The Varieties of Religious Experience,* (first published in 1902) is unsurpassed in its clear objective sympathetic recording of religious conversions. James admits he never personally experienced revelation, but amasses and records the evidence of many others, of different creeds, ages, and doctrinal origins, as well as reporting the affects of such experiences on work and relationships. As a psychologist, he observes not only the persuasive corresponding circumstances and qualities, but also concludes that the only evidence for the existence of God (in this work rechristened the 'field of total consciousness') lies in direct experience and its transformative effects, and not dogmatic or philosophical theism.

He distinguishes four main characteristics appearing in almost all accounts of mystical experience: ineffability (more akin to states of feeling than thinking); noetic (states of knowledge and significance resulting in complete conviction, certainty, and authority); transiency (inasmuch as the intensity is so great it cannot be sustained), and passivity (a sense of being the passive recipient of a gift and often enabled with other gifts, such as speaking in tongues, prophesy, or clairvoyance, and always followed by seemingly explosive and boundless energy).

The genius of science or art parallels some of these qualities, particularly the energy released by vision. It is not that a genius can do things that others cannot, but that a genius can do what many others would be needed to achieve. This burst of energy following revelation finds a natural accord with the instances of inedia, or fasting, reported in saints and contemplatives, many of whom are reported to live without food or drink for years. Rupert Sheldrake in *The Science Delusion*, (2012) quotes many instances of saints, in most centuries, living prolonged lives without food, and without the bodily changes that usually accompany starvation. Normal metabolic changes and waste products are not evident. This suggests that the energy derived from mystical states perhaps taps directly into the energy of consciousness—*prana* or *chi*, long known in mystical traditions in the East—without its conversion to matter, a secondary process.

Religious mystics seem endowed with boundless energy, sleep little, and are remembered by their works: writing, preaching, and travelling. Because the word 'mystic' or 'mystical' was corroded in Victorian times by charlatan associations of séances, table rapping, and bogus emanations of gauze, Underhill adds other distinguishing qualities that describe its essential rigour. It is non-individualistic (not for the glory of the recipient but led by the instinct of love) and avoids structures, either intellectual or institutional. It is active and practical, as exemplified by the concept of the Buddhist bodhisattva who, although released from the wheel of life, voluntarily reincarnates to assist the unenlightened. It is transcendental—aimed at spiritualising the material world, not changing it or disregarding it. And it is both personal as well as intimate, which has often had it accused of being an aberration of human or erotic love, or a psychological inflation that misappropriates religious significance. But that is the view of orthodoxy, which is both renewed by the mystic and threatened by his escape and independence.

Before we draw parallels with the scientific contemplative,

it is important to clarify that while the mystical experience may seem to visit capriciously, its preconditions follow certain patterns, as do its consequences. The path of the seeker is a definite 'way', although it bears different signposts in different traditions—the Cabbala for the Jews, the way of the cross for the Christian, and prior to them, the Eleusinian mysteries, Gnosticism, Zen Buddhism and Sufism. Later emerge the Cathars, Trappists, Quietists, Quakers, and the contemporary monastic sects within each religious family. All have fairly defined periods of fasting, withdrawal, silence, hermetic practices of prayer, and chanting or whirling; all share the intention of quelling the demands of the ego, the flesh, and the interfering mind to prepare or purge for the gift of spirit.

The journey has its stages. Some, like St Paul, were set upon their path unexpectedly on a road to Damascus and captured to follow its call. Just as geniuses of science challenge the prevailing orthodoxy and are often initially shunned, so mystics are usually shunned by the institution of the Church, although later canonized when they are safely unable to speak. Even in more recent times, Père Teilhard de Chardin, both Jesuit and palaeontologist, was, like Galileo, forbidden to publish his equivalent of this thesis, 'The Phenomenon of Man', in which he described the ascent of man as a process of 'Christogenesis'. He also could find no ready words for his experiences and his insights. Drawing from his experiences of both science and mysticism, he invented a bridging concept, and thereby lost the support of both Church and science. Individuals outside of the mainstream are almost always punished until their gifts gain the weight of inescapable truth, which once took centuries, but through accelerating involution, now take decades or years. This is well known, and yet it never changes.

The stages of the mystical path are broadly defined. Firstly, the Way of Purgation (in which the 'givens' of set methods or established answers are shed) is followed by the Path of Liberation, with its bracing wind of uncertainty and cold isolation. Next, is the Way of Illumination, in which the smaller self encounters the larger self, which alone and divorced from the world, thirsts for immersion, for truth, and for reconciliation. This thirst may endure for years until the experience of illumination or spiritual betrothal, the union of two selfs, the *mysterium coniunctio*. The self is absorbed in the greater Self, its true nature to become one with God, or light, or vacuum, or plenum, all beyond words. A period of short-lived ecstasy is then often followed by the Dark Night of the Soul, in which the mystic is bereft of comfort and shorn of society, but also deprived of God, of comfort, or belief. The need for integration stretches the mystic on the cross of conflict, his knowledge newly deep and wide, and life too small to contain it. Only work remains, until the mystic is established in the quietude of certainty, his God within.

This unmistakable pattern, recorded in countless individuals, also shows the involutionary pattern, as though the shoots of isolated mysticism are evidence of the root system penetrating throughout history. From the days of the Pythagorean brotherhood of collective study and discipline, mystical sects were increasingly hounded until individuals took solitary torches in the dark passageways of institutions. Now they exist outside institutions, and cults of many kinds claim mystical origins or ambitions. Conduct, not claims, is the true measure. Following the Dark Night, the mystic is established in a new country of work and has the quiet authority to bestride the material and spiritual without being limited by either, whether wealthy or poor, active or passive. Yet joy and energy seem to dominate, and new creativity arises—often evangelical, for who can fail to want to share good news?

Some brief illustrations in the mystics' own words make the common emotions and knowledge clear, and demonstrate a suspension between melting rapture and clarity of knowledge that has universal validity.

Jelalu d'Din the Sufi poet wrote:

> How great a thing is Love ... for it alone makes all
> that is heavy light, and bears evenly all that is uneven.

Gertrude More:

> To give all for Love is a most sweet bargain.
> O let me love, or not live.

Jacob Boehme:

> I gazed into the heart of things, nature harmonized.
> I recognized God in plants and trees. I come not to
> this meaning through my own reason or my own will
> and purpose, neither have I sought this knowledge ...
> to know anything concerning it. I sought only for the
> heart of God, therein to hide myself.

Plotinus:

> The man is merged with the Supreme, one with it. Only in separation is there duality ... beholder was one with the beheld, God possessed ... a perfect stillness ... the flight of the alone to the Alone.

Finally, from Wordsworth:

> A sense sublime
> Of something far more deeply interfused
> Whose dwelling is the light of setting suns
> And the round ocean, and the living air
> And the blue sky, and the mind of man.

The void in which the mystic is submerged is also the plenum; it is fullness and emptiness, a dazzling obscurity, the barren godhead, and perhaps the un-poetically named 'quantum vacuum' of Laszlo and 'implicate order' of David Bohm. Above all, it is inaccessible to intellect or language, both gift and sometimes, afterwards, curse. Its paradoxical nature is the consequence of the prevailing limits in understanding, applying the intellect to what lies outside its grasp and cannot be defined except by negatives—what it 'is not'.

As a bridging passage between mystical revelation and scientific inspiration, I quote William James's profound understanding of the psyche of a man prepared for either:

> Different individuals present constitutional differences in the matter of width of field. Your great organizing geniuses are men with habitually vast fields of mental vision, in which a whole programme of future operations will appear dotted out at once, the rays shooting far ahead into definite directions of advance. In common people there is never this magnificent inclusive view of a topic. They stumble along feeling their way, as it were from point to point...
>
> As we proceed ...we shall see that what is attained is often an altogether new level of spiritual vitality,

> a relatively heroic level, in which impossible things become possible, and new energies and endurances are shown. The personality is changed, the man is born anew ... (*The Varieties of Religious Experience pp.* 167 & 172)

Innumerable examples are given in Arthur Koestler's *The Act of Creation. (Some quoted below)*

> Jacques Hadamard.
>
> One phenomenon is certain and I can vouch for its absolute certainty ... the sudden and immediate appearance of a solution at the very moment of sudden awakening... (p116)
>
> Karl Friedrich Gauss
>
> At last two days ago, I succeeded, not by dint of painful effort, but so to speak by the grace of God. As a sudden flash of light, the enigma was solved ... For my part I am unable to name the nature of the thread which connected what I previously knew with that which made success possible... I have had my solutions for a long time but I do not yet know how I am to arrive at them. (p 117)
> *'I never think, my thoughts think for me'* (Lamartine p 150)
> *'Our clear concepts are like islands which rise above the ocean of obscure ones.'* (Leibniz p 150)
> *'Man cannot persist long in a conscious state; he must throw himself back into the Unconscious, for his root lives there.'* (Goethe p 151)

Although the patterns of human thought in every discipline echo each other and repeat the patterns of evolution to trace an involutionary path, my essential evidence for this thesis lies in the wealth of mystical experience across countries and eras, as well as in my own small addition to that evidence, which is alluded to in the Afterword; it was the gift that

Appendix

spurred the reappraisal presented in this work. The serpent's invitation suggests that experience alone will furnish proof if enough people accept it and travel through the narrow alley of rebirth, or ego death—which is also the narrow limit of the intellect—into the wider landscapes of consciousness and the experience of love: a love of each, for the self and for all. To the intellect (which is all that these words can reach) love is an abstract; to the experience of consciousness, love is by its very nature warm, embracing, compassionate, and both transcendent and personal. Invariably, its description is shaped by the personality it inhabits; the divine lover no less supreme for filling the longing of man or woman and making each universal.

Recent developments in science are now approaching this inevitability in what is called the 'zero point vacuum'. Nothing reveals the distance between intellect and experience better than the words deemed to describe. There are now areas of psychotherapy that are exploring perinatal experience (Stanislav Grof) and explaining the determinants of manifestation as probabilities or morphic fields (Rupert Sheldrake), stacked one within another, as Kepler's Platonic solids were before them. Transpersonal psychological therapies seek past-life memories, near-death and out-of-body experiences penetrate the world of consciousness, and chaos theorists talk of fractals and holograms. Involution as a collective recovery has exhausted division and rests its case at both ends of the 'all'. The consciousness of man and the whole of creation are now bound together, and in the habitual practice of science, belittled by acronyms—NDEs (Near Death Experiences) OBEs (Out of Body Experiences)—and thereby reduced to characteristics that never suggest transformation and never stop scientists in their tracks to question how these experiences defy or catapult science's certainties.

The experiences that precede scientific revelation share much in common with the circumstances and consequences of inspiration: the moments of recognition. The contemplative scientist withdraws psychologically to a place of purgation (usually after a decade of preparation), and abandons the known paths of solution to a path of liberation. Until submerged by the irreconcilable conflict between alternatives, he turns aside, boards a bus, takes a walk, or dozes by the fire—when suddenly the answer comes. The will and the intellect are in abeyance and inspiration floods in unimpeded. In an instant, months or years of preparation are seeded from some seemingly external source: epiphany or eureka that is sudden, unforeseen, yet curiously recognized as inevitable once it occurs.

Involution

Perplexity turns to immediate lucidity. Conviction in the truth of the answer is unshakeable.

Sir Joseph Thompson wrote, 'it is remarkable that when ideas come in this way they carry the conviction with them and dispose of ideas that were unsatisfactory.' Poincaré talked of 'the absolute certainty which accompanies the inspiration [which] has to be independently verified'. Galton gave a full account:

> In simple geometry I always work with actual or mental lines, in fact I fail to arrive at full conviction that a problem has been taken in by me unless I have contrived to disembarrass it of words ... having arrived at results that are perfectly clear and satisfactory ... when I try to express them in language. ... I have to translate my thoughts into a language that does not run evenly with them. I am obscure through verbal maladroitness not through want of clearness of perception.

Alexander Fleming reminds us that, 'One finds what one is not looking for', and Poincaré that, 'To invent you must think aside'. Lloyd Morgan said, 'Saturate yourself through and through with your subject and wait'. (Koestler. *The Act of Creation* pp145)

We have, throughout this work, also seen the guidance given by aesthetics, and a sense of economy and beauty in mathematical equations, conceptual relationships of harmony, and regularity of shape. Paul Dirac asserted that it was more important to have beauty in his equations, than have them fit the facts. And further, Einstein tells us, 'If the facts do not fit the equations the facts are wrong'.

> Poincaré adds:
> It may be surprising to see emotional sensibility invoked apropos mathematical demonstrations which it would seem can only interest the intellect. ... This would be to forget the feeling of mathematical beauty, the harmony of numbers and form, of geometric elegance ... The useful combinations are precisely the most beautiful. (Koestler p 147)

And Max Planck: 'The pioneer scientist must have a vivid intuitive imagination for new ideas, not generated by deduction, but by artistically creative imagination.' (Koestler p 147)

Bertrand Russell, in considering Pythagoras's allegiance to austerity and the discipline imposed by nature's harmony in number explains:

> Mathematics, rightly viewed, possesses not only truth but supreme beauty, a beauty cold and austere, like that of sculpture, without appeal to any part of our weaker nature, without the gorgeous trappings of painting or music, yet sublimely pure, and capable of stern perfection such as only the greatest art can show. The true spirit of delight, the exaltation, the sense of being more than man, which is the touchstone of the highest excellence, is to be found in mathematics.

For a lifelong non-believer in any religion, Russell comes as close as any thinker to religious rapture, as did Pythagoras at the beginning, before science and religion sundered after Plato and Aristotle, which again, is exactly what involution would expect and predict. The austerity Russell describes mirrors the descriptions of the mystics; the transport may have rapturous delight, but it is essentially, 'Beauty is truth, truth beauty,—that is all/ Ye know on earth and all ye need to know'. Keats's 'Ode on a Grecian Urn' returns the observation to poetry, which is where it meets ordinary sensibility.

From whence comes this recognition of beauty, economy, and elegance if not from our internal experience of being part of it? Why are the standards of elegance so easily recognized by minds familiar with mathematical language: the keys selected by composers, the images of poets? It has guided the genius always. He was not infallible, but if he was mistaken, his solutions tended towards greater economy, greater elegance. Many mathematicians experience almost physical pain at over-elaborate, complex equations that have to circumvent the need for infinity by artificial mathematical devices or by the introduction of meaningless values. Subjective judgement marks the direction for objective truth. Our frameworks of language exclude ourselves, and are always insufficient.

Mathematics may be the closest approach to the mind's expertise, the fine forceps that tease out the underpinning law, but it remains a tool that manipulates and separates; in so doing, it bruises a little by keeping dissolution at bay. René Thom points out that the explanatory power of mathematics declines rapidly as systems become more complex. Yet if this theory proves worthy of mathematical analysis it will be the mathematics we already have, not yet applied to life, which is consciousness. Experience will have to transcend all duality.

The mind's preparedness for these syntheses is often sparked by some so-called chance physical intervention. As a result of the entrenched separation of mind, the synchronicities, or coincidences, in the physical world are shrugged off as anachronisms, improbable perhaps but statistically inevitable. If involution, as the collective re-penetration of memory, was understood, there would be no such thing as chance, for nothing is outside consciousness. The closer to the light of pure consciousness, the more simultaneous thought and event become. The prepared mind may well attract the concurrent event, as a magnet attracts iron filings. What I suggest is that the prepared mind, sparked by seemingly incidental flares, has seeded that new connection; the coordination of all requires the participation of each, and matter obeys consciousness. Coincidence is a word used by those who are asleep to this unity.

Koestler's 'Act of Creation' examines in great detail instances of what he calls bisociation: the sudden, fortuitous synchronicity of previously unrelated fields of thought that reconcile conflict. He calls this 'thinking aside'. Blaise Pascal's ideas on the mathematics of probability likely came from reflections on gambling. Gutenberg's press, with its movable type, was conceived after he watched a wine press and noted the movement of the grapes below the weight, and added that to his knowledge of woodblock carving and the pressed seal. Robert Meyer, who formulated the laws of the conservation of energy, had from boyhood meditated on the transformations of energy in the workings of a watermill. At twenty-six, he accompanied a ship to the Dutch West Indies, where a steersman told him that water was always warmer after a storm. In the tropics, he observed that venous blood was bright, not dark as it is in cooler climates. From biological heat regulation, he derived the laws of the mechanical equivalence of heat.

This hypothesis of a collective process may explain the basis of the zeitgeist: the frequency of seemingly simultaneous breakthroughs, or areas of discovery, in widely distributed people, and the stubborn refusal of the collective to embrace solutions too far ahead of its comprehension. Consciousness appears to pursue its course, with mankind's intellect always in arrears. The waste of good minds broken on the wheel of this refusal can hardly be estimated, since they have disappeared—some into wards of mental hospitals, some feeding the accolades awarded to others. Newton's statement that he could see further because he stood on the shoulders of giants is not always matched by lesser men. Rosalind Franklin's contribution to the Watson and Crick description of the structure of DNA remains known by very few.

The almost simultaneous understanding of evolution by both Darwin and Alfred Russel Wallace was sparked by both reading Malthus's 'Essay on the Principle of Population'. For Darwin, Malthus was the last, definitive locking piece of the puzzle, and the concept of survival of the fittest was the causative link between his wealth of observed and varied characteristics, explaining their transformations, one from another, and their stability once adapted. For Wallace, the visionary, a fever recovered a memory of reading Malthus, which sparked the sudden integration and inspired him to write a paper almost spontaneously.

But affinities had long been seen in many fields—embryology, geology and geographic distribution of similar species—and evolutionary theory had appeared, again almost simultaneously (in Goethe in Germany, Erasmus Darwin in England, and Saint-Hilaire in France) fifteen years before Charles Darwin was born. In other words, Darwin's genius, if genius it was, drew on already established and widespread roots. The rise of organic life, the rise of consciousness, and the relationship of the created world to man needed an urgent all-embracing synthesis. The time for a comprehensive theory was long overdue. 'How stupid not to have thought of that,' was Thomas Huxley's reaction to *On the Origin of Species*. Samuel Butler said, 'Buffon planted, Erasmus Darwin and Lamarck watered, but it was Mr. Darwin who said, "That fruit is ripe, and shook it into his lap."'

The hallmark of genius is a sudden new explosion that connects hitherto unconnected areas of thought. The unification of the observer with the observed, eliminates the barriers caused by the either/or intellect. In *Drinkers of Infinity*, Koestler repeats Socrates' belief that all learning is remembering, which is what this theory of recovery returns us to—no longer surprising. The pupil and teacher are one and the same. The teacher is the unconscious consciousness; the pupil is the conscious intellect. Recognition of what is already known explains the conviction.

For the mystic who has experienced the 'whole', his inability to express this inexpressible whole seldom bothers him; perhaps this is because, within the whole, he has perceived an ultimate purpose—an ultimate meaning in which he is but a small and insignificant, however bounteously enriched, part. Enigmatic economy is quite enough. 'All

Involution

shall be well and all manner of thing shall be well,' wrote Dame Julian of Norwich. Among the geniuses of science and art there are lists of casualties, for madness often follows an inexpressible vision. Rousseau, Nietzsche, Galton, Newton, Robert Meyer, Robert Maxwell, Tasso, Kleist, Holderlin, de Maupassant, Dostoyevsky, van Gogh, Schumann, Hugo Wolf, Georg Cantor, Wolfgang Pauli—the list is endless. Perhaps these men, having drunk from truth, were afire with a thirst that could find no assuaging. Genius is the preparedness to pit oneself recklessly against accepted beliefs, and it occurs in science and art, as it does in religion, at times of crisis. It gets us out of a corner by blowing up the building but it often takes the inspired's sanity with it.

Through involution, science has, like the mystical traditions, brought mankind to the doors of both perception and transcendence, but only the individual can bow down to enter. The serpent affords access to the shared 'inwardness' of all creation, with which each of us is united. The inner light is the union with all, through which we revalue our uniqueness, and that of each other. All is lit from within.

This can only be validated by direct experience. Although this work points towards an exit from the ordinary ego-bound world, it cannot not take you within this alternative or find reason to elaborate the experiences from which this creation sprang. For those already steeped in such encounters (through past-life therapies or psychedelic exploration) that would add little to what was already known. Instead, this book is addressed to the disappointed, to those who search for understanding of the paradox whereby the glory of the created world—its enlargement through the capacities of its art, its worship, and its science—is no longer enough. The deep malaise is the prelude to a new glory; this book attempts to explain how we arrived at malaise, why now, and wherein renewal might find its love and its light. Closing doors is always the precondition of opening new ones, or finding them graciously opened when approached.

This theory of involution has waited forty years to find a receptive scientific audience. I can't help but feel we are surrounded by ripe fruit, but that, against the hold upon it by entrenched scientism, no amount of shaking will bring it down.

Afterword

**Soul suggests Reason should explain herself…
About the Author's Experiences underpinning this Thesis**

*Why so tardy were you summoned to express
The wax conviction that has cradled you through life?*

How can it bring more ridicule?
To carve the slab of science with a poet's pen
Than to pretend deduction or the logic net
Knotted piecemeal from ad hoc experiment?

When I have heard a music stripped of word?
Consumed a knowledge boned of mind?
Worn fabric smooth with neither warp nor weft?
A stomacher of synthesis
Laced through with gold alternatives
To argument? (That eye you opened on that beach
And never closed completely…)
Though I attempted to forget the obligations it imposed.

*Conviction? Sense? You are unwise
When Schopenhauer, Hegel, many, came before
With better minds than yours, and as ignored…
The Perennial Philosophy never wanted
Advocates. Why yours?*

*You mention obligation. What did it garner in
To drive your life to this last moment's pause?
Why so uncertain of your suit? When you already know*

Involution

The thickets of rebuke. What will you add
To that broad highway of precarious thought?

Why else did you bequeath me solitude
From birth? But to prepare my ear
To hear a different music? At each turn the way
Moved further from collective certainties
Towards the single moment when
That eye removed all doubt of what I should and was.

Why birth me like a small papoose
Among the larvae in their single cots
The only white skin nestled amongst black?
Why take my father from me, ere I weaned?
Or set my sandaled feet towards a school
So I could shape a language to reject?

All down the corridors of years behind
I was instructed carefully. Each discipline
A stone to set, to counter with, or craft
A glove to fit the hand that measured it.
When stripped away it peeled my skin
Printed with the mirror inside out.

It is not thought that brings me to this point
But perhaps a hope that science has run dry;
Its hypotheses are vacant and its dearth
Betrays the limits of its intellect.
Because I know I know, I must believe
That old words now might meet new tympani.

Make method now your master; enough of this
Slow measure. Strip out persuasion and apology
There will be those who choose to understand
For they already are.
For the others, who will read debatingly,
To evaluate the wellsprings of your views, a tongue
Dissolves the sherbet of this fresh new taste

Afterword

Upon a palate furred with doubt…
I suggest you introduce yourself with politesse
Why should you expect an audience rapt
When they are new to this?

Well, that eye, that lucid green gold lens
Appeared, quite uninvited. I cannot account for how
It found me waiting on a beach, sideways seated with a man.

It came to me much later that the Eye
Was both a hinge and trapdoor. The first
Pinched out my life before, the second sealed return.

But on the occasion it appeared, the dawn sweet sweep of sea,
The hiss of surf, the soft raiment of caressing wind
Was fitting bed on which to focus that all-seeing lens.

Before its birth, it drew aside the curtain of the 'real'
By an incision in the landscape; shattered spikes of surface glass
Slid on lasers sideways, and completely disappeared.

It was as though the waxwing had been stunned and as it fell
It pulled the painted cheesecloth down. The landscape was no more.
Gone were four dimensions with all newly dissolved walls.

The luminescence was more atmosphere than light; more
Clarity transparent, a boundless nothing of the all,
Pregnant with potential birth, yet already born.

Leaving the loom unoccupied by either thread or time
Into which that all-seeing Eye swelled slowly into sight…
To gaze behind and forward, clear and compassionate.

Steady and unblinking it remained, while we perused
Its green-gold iris ringed and flecked… Nor quivered
While we spoke. From either side of nothing…remained impervious.

Involution

The only thing that was, that Eye, all visual else dissolved,
Though words exchanged from his mind to mine were spoken through the air…
Until confirmed identical. It slowly disappeared.

From the periphery, the wings of stage, the landscape now returned,
The spicules of the curtain rent, moved back to reconnect…
A face with stricken swimming eyes, a freshly pearling sky…

Plato's surface of that peerless cave, of false persuasive sense.

(Aside)
(Reader I would draw attention to the management of two
Corroborating witnesses. Most encounters are dismissed
Through the single error, as inflation of crude faith…
That icon Eye is everywhere, above countless triptychs painted,
Kohl black and stark with staring, to find one returning gaze.

Here on an unlikely beach, where two adulterers explored,
With time for talk and languid…there could be no mistaking
What was never sought.
Nor gloaming shadow to obscure, debate a smoked belief…
Instead it was Excalibur, buckling up a breastplate for an internal strife.)

A moment, wait, you anticipate, we all need some measure…
Had there been no bobbing rabbit bait? No prior prescience
To lead you down from dappled light
Towards a cavern, vast, sidereal,
No intimations of that clumsy word, of immortality?

Yes, now you ask…a few weeks earlier, with hash-heads that
I hardly knew, cross legged, around candles and those
Curling incense sticks. I was far too sober, at a loss

How best to pass an evening in such dreadlocked company?
I could not read (the light was low)
Devoid of words, they seemed benign; smiling, vacant, lost.

Afterword

I had drained a glass or two, the evening meal washed up:
I cushioned back to give myself to reverie, or sleep…
Instead I was imprisoned in a meningic bulldog clip.

A rubber cap that creased the brow, a skin integument
That squeezed my eyes below the skull. The lamellae
Of my mushroom brain clamoured to escape.

If I could only puncture it, and splinter through the bone
I might escape, not just the room, but
All confining frames.

Giddy now, I took up thoughts, unsorted skeins of wool,
To spiral down, full queasy, to the vortex in my chest,
My heart a burning coal that charred the basket of my ribs.

Then exploding, as a bullet might from the barrel of a gun,
I floated free, above my husk, lying there, oblivious…
The candles flickered. One fell over, guttered, and went out.

I rose upwards, still compelled by some escaping breath…
Behind, a rope uncoiled itself, a sheen umbilicus
Twisted like liana rope, half violet, half blue…

It pulled from my rear occiput, to the narthex of my chest,
The solar plexus held the string of its full inflating kite
That billowed in a larger space, than the confines of that skin.

I rose above that seated group compelled snakelike to slough
The cell of that dry corpse. Through the roof I sailed serene,
A barrier of density, that's all, and out into the night.

Above the arms of Biscayne Bay, the lights along the shore
Held pearls against the moon streaked beach; the palms below fanned gills,
Dipping acquiescent hydra heads, sifting black sea ink of night.

Involution

Through the transparent roof below, foreshortened figures sat;
Their crossed patella cheeks in shorts, rouged in flickering light,
Their gothic immobility needed Escher's pen.

That was below. But harking up through sonic spheres
I was trumpeted to heaven. The stars a bridal net of light
Cast to marry lover-fish with thirsty poets, tramps.

A thought sliced through the marvelling. I was unfettered, free…
The cathedral of all thought was mine, including his and yours.
In the chancel, crypt, or vestry store, I could travel anywhere

Through space or time, unbidden, but below the radar mind
Africa, my mother, I'd take her a gift…appear.
I somehow felt a single thought would betray my visit there.

A mighty wind rose in my ears, blew out the landscape bay,
The lights, the stars, and flickering thought. Blake crayoned shoulders
Setting south, took leave of corpse below.

A second thought undid the first. How would I come back?
The landscapes of my children's breath, their honeyed arms, their tears?
Who could be sure the distance at which the string might snap?

One small apprehension did the thin thread twitch.
The question's unwise doubt yanked a cannon from a cliff,
To plummet down with folded wings, a gannet after fish.

Slammed headfirst into solid rock, the arid brittle shell
Quivered in bleak ecstasy to reclaim my sick return…
Drummed heels upon the floorboards, relocked my ball and chain.

In the reverberating shock a Richter scale could measure
A voice behind me murmured
'Wherever have you been?'

*

Afterword

(No doubt the reader has observed that she and I both use
Similar words and images. How could we otherwise?
Her brain is soul's receiver; I make use of what I find.
While she dreams I raid the attics and the foetid slimy cellar
To furnish further linkages, enrich the store of images
Drawn from more recent life.
Hers has been haphazard, but yields rudimentary tools
I make fit for use.
I admit I did insinuate
She should set this rough-cut jewel in metallic verse.

For years I have observed her collections of soft moss
From other theory's nests; like those sharp- shooting Corvidae
That posse in to steal the twigs from settled homestead folk…
She has sweated in their kitchens…
Speared nimble facts on forks…
Creamed all research with pepper sauce
Judged unseemly by the tonsured with a taste for splitting hairs…
You cannot hatch out fertile eggs on other's arguments.)

*

So let us glean and clear the mind on which that eye then opened.
What had your escape led you to understand?
Did it leave clear footprints in your ward of wife?
Dogs body, nurse and cook…forsooth, a pebble in the hand?
What could you keep, what learn, from your single flighted trip?

At first I wept small tears of gall that I had so mistook
What yogis search for all their lives, and anchorites on pillars:
I had stubbed my toe on gold, and left the impress of my teeth.

Later though, the fork of thought dug over all belief:
The brain's no seat of governance, the eye no picture scanned;
Both were full the opposite of what I thought I'd learned.

Involution

We are but projectionists in the cinema of life
Screening clichéd classics (and some in black and white)
Instead we could create a new, if we turned up the lights.

All the pictures shown are those that we agree,
The programme arbitrator screens collective memory,
We all nod our heads through genetic symmetry.

The anchorage to energy, gravitation claims
Everything in synthesis, binds all to coalesce.
Dark matter is invisible, it is the unthought thought.

It passes through all objects, as I passed through the roof,
Invisible, timeless, Ariel girdles the earth,
In less time than a photon flies, the right thought can ignite.

Or fall back like Icarus dissolving feet of clay.
Torn from collective melting pot, an idea consents to fly,
Only if fine sanded smooth for the mortise ready made.

This was your conclusion before the summit's sharp ascent?
On your return, you packed, ready to roll your tent
In pursuit of yogis to the oxygen-spare heights?
Certain you could breath such air, avoiding
Crucifix or spear?

You mock me (there are precedents) but full well you know
I did nothing like. My world was newly inside out,
Straddling ice-floes drifting apart, irreconcilable demands
Of children at my dirty knees, my head in unhelpful cloud…

It took my life to reconcile
(It's what this tale conveys…)
Your prologue shaped my epilogue…
The harvest of a quiet eye
Has just been brought to bed.

Endnotes

Footnotes to Canto 1.

1   Blake's famous painting *The Ancient of Days* portrays God parting the cloud with a pair of compasses.

2   Siddhartha Gautama (Buddha) was born in approximately 500BC of a wealthy family. At the age of twenty nine he abandoned his privileges and took to a life of abstemious renunciation in search of understanding about the suffering and impermanence of worldly affairs. After a period of self-imposed privation followed by extreme asceticism he adopted what he called 'The Middle Way' of neither ascetic extremes nor indulgence. The founder of Buddhism, he experienced enlightenment and thereafter spent his life with disciples and teaching the Dharma.

3   The coming of the Messiah. Jesus Christ is, to the Christian tradition, the only Messiah but to other traditions the messiah has come in many different incarnations. Messiah literally means 'The Anointed One', the messenger of God. That is the interpretation implied here.

4   The parable of the talents (Matthew 25:20): 'And so he that had received five talents, came and brought other five talents saying "Lord thou deliveredst unto me five talents; behold I have gained besides them five talents more". (v21) His Lord said unto him "Well done thou good and faithful servant; thou hadst been faithful over a few things, I will make thee ruler over many things…"

5   Casting pearls before swine (Matthew 7:6): 'Give not that which is holy unto the dogs, neither cast ye your pearls before swine, lest they trample them under their feet, and turn again and rend you.'

Involution

6   The Qur'an (Recitation) is the holy book of Islam, believed to have been dictated to Mohammed by the Angel Gabriel. It sets out the essential practices of Islam and the rules governing right conduct. Considered the most beautiful of poetic Arabic, it set the standard for later Arabic literature and the basis of Muslim education lies in the complex levels of interpretation it permits. Reciting the Qur'an in Arabic is considered the supreme achievement of an Islamic scholar. 'Brick like' refers to its influence and power, in defined guidance and importance, not its appearance, often very ornate. Any destruction of this literal word of Allah is considered sacrilegious. It assumes an acquaintance with both Jewish and Christian history and doctrines, mentioning Moses as a prophet and Jesus and Mary, the latter more often than the Bible.

7   Homer's *Odyssey* recounts the return of the heroes of the Achaean army to the parts of Greece from which they had been gathered and particularly of Odysseus. Its events are dated later than the war recounted in the *Iliad*. The word is now commonly used for any long journey that has an element of personal quest or extended development, metaphoric as well as physical.

8   Very little is known about Homer, believed to have existed somewhere between 1334 and 1150BC. Herodotus puts him much later at about 850BC, which is supported by an analysis of his knowledge of the events he recounted. He is thought to have been born near the coast of Asia Minor and therefore an Asiatic Greek; he refers to the men of Greece as Achaeans (or Danaans or Argives, not Graikoi). Whether he could write, or whether his works were derived from a compilation later of the oral traditions of the sack of Troy is uncertain. Clearly in writing about the tenth year of the siege, with a cast of characters he knew well, the story suggests a long, possibly oral tradition.

9   Paris, sometimes referred to as Alexandros in the Iliad, was the son of Priam, King of Troy. He had seduced Helen, the wife of Menelaus, and taken her with her full consent from Sparta to Troy. This was the event that led to the Trojan War, Agamemnon's attempt to recapture his brother's wife. As a character he was renowned for his lust, a quality he was granted by the Gods. He is described by Homer thus:-

Footnotes to Canto 1

'Evil Paris, beautiful, woman crazy, cajoling,
Better you had never been born, or killed unwedded
Surely now the flowing haired Achaeans laugh at us
Thinking you are our bravest champion, only because your
Looks are handsome, but there is no strength in your heart, no courage…'

He brought judgement on Troy through his intemperance regarding Helen.

10   The Trojan horse was constructed by the Achaeans as a means of penetrating the ramparts of Troy. They filled it with soldiers and left it outside, calculating that the Trojans would wheel it in through the gates. It does not appear in Homer's narrative of the Iliad but is mentioned in the later Odyssey and is the episode best known about the Trojan War.

11   Mycenae, the citadel of Atreus (and his successor Agamemnon), was fully excavated by Schliemann in 1841. He took the events recounted in Homer as historically factual and uncovered the lion gate and the burial places which he attributed to Agamemnon and Clytemnestra. Olympus was the mountain on which the Greek Gods were thought to reside.

12   This is a reference to John Keats's 'Ode to a Nightingale'. 'The voice I hear this passing night was heard / In ancient days by emperor and clown: / Perhaps the self-same song that found a path / Through the sad heart of Ruth, when, sick for home, / She stood in tears amid the alien corn…'

   This poet is suggesting that just as the Messiah repeatedly came to arrest the forgetfulness of God, so too poets, in different climes and times, appeared to re-evoke the religious or spiritual awareness of those alive to their messages.

13   Dante Alighieri (AD1265-1321) was a Florentine whose *Divine Comedy* recounts his soul's journey in the company of his guide, Virgil (70-19BC), portrayed as his mentor and protector. The first section *The Inferno* takes them through hell where they meet all those condemned for different crimes, (one of the most heinous in Dante's book was the

crime of indifference- a thought for our day). The second section, *Il Purgatorio*, recounts meetings with the repentant whose currency is love (as opposed to the hatred of the Inferno) through which sin is gradually shed and the soul prepared to enter paradise.

In the final section, *Il Paradiso*, the soul ascends to paradise and the circle of the soul's travel away from God is completed by its return. Almost immediately he encounters a vision of Beatrice and asks her how it is possible for him to climb up through the lighter elements he encounters. She answers 'Everything that is created / Is part of a mutual order, and that is the shape / Which makes the Universe resemble God'. In this experience Dante echoes the experience of Buddha in nirvana. Beatrice explains, 'You should wonder no more / At the way you came up, than a river does / At the way it goes down a mountain to the bottom. It would be a marvel if, without / Any impediment, you had settled below / Just as it would if a flame stayed on the ground.'

In other words the return to God is the very nature of the soul and the completion of the journey. The remainder of his journey is in the company of Beatrice who represents natural love and replaces Virgil, the guide of intellect and understanding. This necessity to let go of mind in order to enter a union with God (or love) echoes the Buddhist contemplative, 'middle way' approach to abandoning the separate existence of human attachment to anything, even ideas.

14   Virgil's *Georgics* showed him to be both knowledgeable and practised in the arts of husbandry. His tender knowledge of agriculture, birds, trees and bee keeping is astute and experienced, observant and reverent, inasmuch as the divine connection with the earth's riches are celebrated; Pan, the 'master of flocks', Minerva who first 'discovered the olive', the seasons governed by the planets... 'Amid the slow months, a gap is revealed between Virgo and Scorpio'. All of nature and man's stewardship of it is celebrated as an entrusted covenant between man and the Gods. It is difficult to restrain admiration of Virgil's deep humanity, as well as his knowledge.

'Vines there are, one suited to heavy soil, one to light soil;
Psythian are best for raisin wine, while thin Lagean

Footnotes to Canto 1

On its day will trip your feet, and tie your tongue in a knot.
There's the purple grape and the early. What poem can do justice
To Rhaetian? Yet even this cannot compete with Falerian.
Amminean vines afford us the most full bodied wine,
To which must yield the Tmolian and even the royal Chian....

There is no end to his knowledge of the preferences of trees:

'Willows grow by streams, alders in soggy marshlands;
The barren rowan tree is found on rocky hillsides
Myrtles crowd to the sea coast...
Vines love an open hill, yews a cold northerly aspect...
    And those grown elsewhere... India alone
Gives ebony, Arabia the tree of frankincense....
The soft cotton that glimmers in the plantations of Ethiopia?
The way the Chinese comb the delicate silk for their leaves...'

15   Dante was the first educated man to use the vernacular Italian, rather than Latin for his greatest work. It is, throughout, set in three line verses adopting a pattern of regular rhyme (aba, bcb, cdc, ded etc). That is the meaning of 'terza rima' though it does not easily lend itself to such a pattern in English translation.

16   Milton's *Paradise Lost* is another in the great canon of poetic odysseys, in which the devil now has the best part, as does Mephistopheles in Goethe's *Faust*. Faust is tempted by Mephistopheles who engages in metaphysical jousts, but like Beatrice awaiting Dante on the heights of paradise, so Faust is awaited by Gretchen.
In these later poetic odysseys man is represented as having fallen away from a consciousness of the divine. The poet John Milton was blind, as was Homer reputed to be, which gives some additional merit to the supposition of poets reincarnated to arouse man to listen rather than to look.

17   Quidditch is the aerial game played on broomsticks in the Harry Potter novels, in which a winged 'golden snitch' is the precious target and only captured by the 'seeker'. All the other players score goals but the real prize is the capture of the snitch which remembers the person who first captured it. Beaters aim to prevent the seeker's encounter

Involution

with the elusive snitch. Capturing the snitch ends the game.

18   In *The Divine Comedy* Dante draws diagrams of the structure of the ascent to paradise. The Mappa Mundi placed Jerusalem at the centre of the inhabited world of the Northern Hemisphere. The Inferno is depicted as an inverted narrowing cone with the lesser sins (the ordinary venalities, such as gluttony, promiscuity, and avarice) punished on its wider higher levels and traitors (to country, family or causes) in the deepest darkest levels nearest to Satan. Mount Purgatory is placed opposite Jerusalem in the Southern Hemisphere as an ascending mountainous cone of spiralling levels up cornices towards earthly paradise (Eden). Each level has its own repentants; the lowest were the excommunicated and apathetic, followed by the seven mortal sins of pride, envy, anger, sloth, avarice, gluttony and promiscuity. Dante is surprisingly forgiving of sins the church would have ranked differently, but reserves his greatest condemnation for mental betrayal, indifference and treason.

True paradise has no 'form' represented but is described as 'And suddenly, it seemed that, to full daylight/ Full daylight had been added/ As if he who can had put another sun in the sky.' It is clear that true paradise is a state of being without form or structure.

19   Aristotle's rigid hierarchy of regular planetary motion in perfect circles is therefore reflected in Dante's conceptions of levels of sin and repentance, with the sun (Apollo) the centre of all creation, to which all aspire to ascend.

20   Iambus is Greek for 'lame', and iambic pentameter describes a metrical foot with two syllables, a weak step before a strong one. In iambic pentameter there are five iambs to each line in regular stresses (believed to be the natural rhythm of language… as though stepping or walking). It is well illustrated in Grey's 'Elegy Written in a Country Churchyard…
The cúr/few tólls/ the knéll'/ of pár/ting dáy
The lów/ing hérd/ wind slów/ly ó'er /the léa.

The pterodactyl is an extinct winged reptile.

Footnotes to Canto 1

21   In the *Georgics* Virgil gives the remedy for the death of a hive:-
'If a man's swarm shall suddenly fail him, so that he has no source for another brood
It is time to detail the famous invention of an Arcadian
Bee-master; the process by which he often made
A culture of bees from the putrid blood of slaughtered bullocks.
I'll give you a full account….

First a small place is chosen…; they close it with a pantile roof
And prisoning walls…
A two year old calf is obtained, whose horns are beginning to curve
From his forehead. They stopper up, though he struggle wildly, his two
Nostrils and breathing mouth, and they beat him to death with blows
That pound his flesh to a pulp but leave the hide intact.
Battened down in that narrow room they leave him, under his ribs
Laying fresh cassia and thyme and broken branches…

Meanwhile, within the marrowy bones of the calf, the humours
Grow warm, ferment, till appear creatures miraculous—
Limbless at first, but soon they fidget, their wings vibrate
And more, more they sip, and drink the delicate air:
At last they come pouring out, like a shower from summer clouds,
Or thick and fast as arrows…'

22   Worker bees, although nominally female, are sterile. The destruction of a hive is common as soon as the queen bee dies, with the hive being newly established elsewhere. More often she will be replaced by a larva, prepared by being fed royal jelly, until ready for the nuptial flight during which she is fertilised by a drone or drones (which then die) and so becomes the new queen.

23   Richard Dawkins's book, *The Selfish Gene,* gives ascendancy to the role of genes as the dominant driver of evolution. For him the growth of the organism, in which the genes are carried, becomes almost secondary; its purpose being to secure the genes' survival through the competitive survival of the organism, with any advantages it might have over other organisms being selected for by 'natural selection'. This extends classical Darwinism (which focused on the organism's selective advantages over competition). When *The Origin of Species* was published, DNA was

not understood but the dominant role of genes (and the mechanisms for genetic determination) are now central to neo-Darwinism, with organisms being almost secondary 'packaging' in which genes survive, duplicate, and get distributed.

From small variations within a population, Darwin proposed that slow changes through evolutionary time would lead to new species arising. Since the understanding of the structure of the DNA molecule by Watson and Crick in 1956 it is known that small random changes to the DNA molecule (by mutation— unexpected copying errors) might lead to such variation. This slow incremental process relying on random events (most of which are deleterious) are thought, by many, to be too slow a process to explain the great complexity of both biological structures and behaviour.

First the random change has to happen (radiation or chemical causes are most common), then that change has to be among the few that get the timing right so as to endow an adaptive advantage, which natural selection then favours with both survival and breeding. Variation is also questioned as insufficient to explain speciation, since species are historically defined as intra-breeding populations. Most new inter-species hybrids are sterile. It is not seriously questioned that survival of the fittest explains the gradual improvement of a species; rather what causes problems is the 'arrival of the fittest' the new jump to an entirely different species. Darwin acknowledged that natural selection was 'niggard in innovation'. It is reactive, not pro-active.

Yet the DNA molecule that codes for all species has retained the records of change, and must therefore itself have been changed over time. What triggered those changes? Conventional neo-Darwinism allows the DNA to dictate the changes in the structure and habits of organisms, but the reverse, (commonly called Lamarkianism; 'The Inheritance of Acquired Characters') does not allow for the habits of organisms to change the DNA, or be inherited by succeeding generations. This is the central precept questioned in the Theory of Involution. Recently, changes to DNA have been shown due to toxic exposure and 'epigenetic inheritance' over successive generations indicates that factors other than genes influence inheritance.

24   Kinder, Kirche und Kuchen, the three 'K's (Children, Church, and Cooking) were the roles defined as suitable for the real Hausfrau of German ideals; endorsed by Hitler's exhortations for wives to stay

at home to increase and safeguard the Aryan population. Breeding was the purpose of defining restricted roles.

25 Karl von Frisch first interpreted the 'waggle tail' dance of the bee when it returned to the hive after foraging at some distance away. If the food source is close to the hive it just performs the 'round dance' which excites the others who rush off in the close proximity. If the food source is further away it performs the waggle tail dance, a kind of figure-of-eight with a straight run in the middle. The direction of this straight run tells the compass direction of the food: the energy and speed of the dance indicates the distance. Since the dance is performed in the darkness of the hive, the other bees feel the dance rather than see it, and the performer also accompanies the dance by emitting piping sounds.

But, and here is the aspect of the miraculous, the compass direction has to be interpreted horizontally by the other bees. Bees can perceive polarised light, so even when there is cloud the bees know the position of the sun. Von Frisch discovered that bees have an internal 'clock' that tracks the movement of the sun across the sky, even when they have been confined in the hive for hours. The angle of the straight 'run' of the dance gradually moves, like the hands of a clock. In the Southern Hemisphere the hands of the bee clock move in the opposite direction!

Now the dance code is exactly as conventional as a map pinned to a wall. Straight up means 'in the direction of the sun'; straight down means 'in the direction opposite to the sun'. All intermediate angles are in relation to this vertical axis: 50 degrees to the left signifies 50 degrees to the left of the sun's direction and so forth.

The rate of turning (and waggling its tail) through the circles indicates the distance; the closer the food, the faster the rate.

Like ocean navigators, bees have approximately eight compass points like the quadrants N, NE, E, SE, S, SW, W, and NW. The bees, who detect the kind of food on the returning bee laden with pollen and nectar, burst out of the hive and carry with them the compass direction and distance they have learned must be followed.

Given that this extraordinary dance must have evolved over time, Dawkins postulates that the bee ancestor, living in the open and able to see the sun, would simply have offloaded her cargo and returned to the source of food. Because the sun would be visible and the bee was outside the hive there was no need to develop the convention of the vertical axis of the dance. He suggests that it evolved from repetitions of the take off 'run' directly back towards the food, which through repetition persuaded other bees to follow, as a ritualised form of emphasis. The figure of eight would 'balance' the direction of the run itself: if there was only one way adopted for the turn, the 'run' could indicate either direction (at the end of the run). By turning first left and then right the figure of eight evolved to isolate the accurate direction.

The longer the distance from food, the slower the dance, would be the natural consequence of fatigue in the returning bee. So far, so plausible.

The next step, to explain the adaptation to the vertical surface, Dawkins accounts for by suggesting that the insect nervous system balances orientation to a light source, and that there seems to be a neurological transfer between this visual stimulus and the gravitational sense. Beetles that maintain a 30 degree angle to a light source on a horizontal plane will 'transfer' this angle to the vertical when tilted on a board vertically and deprived of light. (Continued in Note 26).

26   This was one of Dawkins's answers to the 'argument from incredulity' which finds the sophistication and intricacy of this language of the bees a justification for questioning a slow evolutionary cumulative adaptation. Although each stage of the plausible explanation makes sense, it requires not only the dancing bee to adapt to darkness and the vertical orientation of domestic bees, but all the others to substitute vibration and audition for vision. It is a stark example of what I would quote as the 'Nothing Buttery' (the term coined by C.S. Lewis to describe science's reductions of intricate events to being 'nothing but' arrangements of components) of neo-Darwinism. Each element of the bees' complex communication is isolated and accounted for, while ignoring the integrated matrix required to adopt and interpret the initially experimental, or almost accidental, until it becomes fixed and ritualised in a population.

Something is missing; perhaps that something is the relationship of bees to flowers, the extended ground of the bees' being, whereby both flowers needing to be fertilised, and bees needing to use nectar had a part to play in ensuring the benefit of both, by a language serving both. This is revealed by the preference of flowers reliant on bees to

Footnotes to Canto 1

be prone to the ultraviolet spectrum, which bees can see. Even if they appear white to us, they are often patterned in ultraviolet, which we cannot see but which send signals to bees. This is, I believe, the consequence of isolating one organism and its gene pool, and studying its relationship to time, rather than seeing the co-operative nature of all genes in collaboration. 'Competition' remains the central tenet of evolutionary explanations, remaining even when the continuity of the genes 'self interest' has replaced the organism's 'self interest'. Imitation and communication within a hive could explain this sophistication far more easily.

If genes are now understood to collaborate with other genes in their joint endeavours to ensure survival within the organism, then perhaps what Dawkins calls 'arms races' (the competitive adaptation between species) could be modified somewhat for the understanding of the perfection of all creation. He should read more Virgil perhaps, although he would become apoplectic at the story of 'spontaneous generation' of bees within the putrid flesh of a calf. Nothing I have indicated disputes the postulates of Darwinism, merely probes their possible extension into the gene pool of collective consciousness.

27   Quarks. The name derives from James Joyce's 'Finnegan's Wake'. What were once called cosmic rays were found, in collisions, to give off showers of particles that were not the already known particles of proton, neutron, electron and neutrino, but entirely different and more fundamental components of the protons and neutrons in which they were combined; in twos as in mesons and in threes in heavier particles still. Through particle accelerators which magnetically increase the speed of collisions a panoply of these fundamental particles has been 'found'. None have been seen but their differing properties are described as spin (in one of two directions) and 'flavours' (five in number, one used to be given a 'strangeness' value) and 'colour' of which there are three. Spin, flavour and colour combinations produce some thirty possibilities like a bag of jelly beans.

The interactions between these quarks seem to adhere to concepts of symmetry, which Hoyle suggests indicates that they simply adhere to our concepts of space and time. This might imply that space and time are not fundamental concepts, but merely the reflection of the imposition of our limited experience and placing causality 'behind us'. Hoyle's supposition is supported by the reversibility in time of the equations of Maxwell which work equally well in both directions. 'Advanced waves' from the 'virtual future' influence the events of the present and offer the alternative concept of a positron as merely an electron moving 'backwards in time'. It is here that the naming of names (where space-time replaces space and time) shows the deficiency of both language and mathematics to reveal anything other then the processes of mind. The behaviour of quarks defies common sense. Neutrinos pass undetected through matter, and do not interact with it at all, so tracing their existence requires them to be artificially 'produced'. Do they come into existence merely because they have been conceived collectively by physicist's minds?

Could it be that quarks are the variants of mind, and once conceived by the collective mind of particle physicists they are collectively created to adhere to the necessary symmetry?

28   Narcissus fell in love with his own reflection as a penalty for rejecting the advances of Echo, and was condemned to mistake the imaginary as real.

29   The early Kibbutzim in Israel held all property in common and children were raised in collective nurseries. It was thought this would make them proto communists willing to fight for the collective.

30   Richard Dawkins held the Chair at Oxford created for the Charles Simonyi Professor of the Public Understanding of Science. He has recently become a vociferous opponent of religion in all its forms. This is given vehement expression in his book *The God Delusion*.

31   'Pupillary poetry' was a term defined by C.S. Lewis as the poetic imagery scientists use to assist their thinking. At its most powerful, in the right mind, such as Kekulé seeing the benzene ring as a snake devouring its own tail, or Einstein imagining riding a light beam, or Faraday's 'lines of force' which assisted his thinking on electromagnetism, it can generate creative breakthroughs. 'Bad poetry' is what Dawkins refers to when metaphors are thought to have intrinsic truth, or superficial resemblances to other things which endows them with meaningless meaning. This entire book will be considered prone to the charge of bad poetry.

32   Random error through mutation is the central means by which it is thought genes slowly alter through time, and on which evolution depends for its advances. Under the influence of heat or chemicals, errors in the genes' replication (the genotype) confer occasional advantages to the organism (the phenotype). Mostly they are deleterious. Any advantageous random changes are preserved by natural selection 'favouring' the survival of the 'fitter'. *The Blind Watchmaker* is the title of a book by Dawkins in which he seeks to demonstrate the improvement of species by this mechanism. The title is derived from a metaphor used by William Paley to demonstrate that if a watch was found by someone who had never seen one, its intricate cogs and wheels would imply 'design by intelligence' and for some purpose. Nature is so intricately interconnected, and its mechanisms so perfected, it gives rise to the supposition of deliberate design and therefore the 'Designer God' that created it. Dawkins seeks to dispel any such conclusions by showing that blind chance and a great deal of time explains everything. The 'blind watchmaker' of natural selection simply 'weeded out' the less adapted, leaving the fitter to succeed and improve. Involution simply places the 'fitness' within the organism instead, a product of mind, giving natural selection a subordinate and secondary role.

33   Originally the use of genotype and phenotype distinguished between the genetic 'blueprint' in the reproductive cells (the genotype) and the manifestation of that blueprint in the characteristics of the organism (the phenotype). After *The Selfish Gene* placed the emphasis upon the evolutionary self-interest of the gene in fostering its own survival (rather than the survival of the creature carrying it) the concept of *The Extended Phenotype* increased the genes' spheres of influence to the environment selected and the habits of the organism. The extended phenotype therefore includes the parasites, the symbiotic relationships, and the realm in which an organism functions. This can even include other creatures geographically distant, as would be encountered by migrating birds.

34   The complexity of the vertebrate eye is, by many, thought to defy explanation by a process of slow refinement over time because the partial development of one aspect would not, on its own, confer the advantage that would assist survival. Dawkins, on the basis of

computer modelling, has claimed that the evolutionary time available would have enabled the vertebrate eye to arise four times from scratch. He also argues that the improved design in the octopus (where the retina is reversed in relation to the nerves carrying the retinal impulses) shows, through the relative imperfection of other eyes, the evidence of slow and not always optimal (pun deliberate) methods of development. In short the vertebrate eye is imperfect, and hence not 'designed'.

35   Einstein famously said 'I do not believe God plays dice with the universe' in response to the hypotheses of quantum mechanics which he could never wholly accept, however neatly they squared the circle. Quantum mechanics and relativity co-exist but remain conceptually irreconcilable.

36   This refers to the argument that the suggestion of blind chance as the mechanism of evolutionary advance was tantamount to suggesting that a tribe of monkeys, randomly hammering on typewriters, would by accident, over enough time, produce a line or more of Shakespeare.

37   The chemical responses that take place when emotions are involved (in sympathy or self-sacrificial or humorous situations) severely impede the adrenalin needed for flight or fight. It is well known that laughter inhibits sex. So how did these 'sympathetic' nervous and chemical reactions arise when their presence would impede the chance of survival? The field of animal behaviour is just beginning to get to grips with evidence that these so called 'human' qualities are present in lower mammals, and that they confer, through mutuality, a better likelihood of survival. Group selection has recently come to the fore to explain these aspects.

38   Tears and the evolution of the tear ducts have no satisfactory evolutionary explanations. It has recently been shown by de Waal that many species of animal are commonly shown to have empathy, not only with one another, but with other species, even sacrificing food to protect another from shock in experiments, or sharing food voluntarily. Increasingly, the higher emotions, thought the unique province of human kind, are now being seen in earlier evolved mammals like dogs, great apes, in dolphins and perhaps even in birds. It could be

that the influence of man's co-operative consciousness has moderated the emotions of animals, but the convention of time moving only in one direction from 'lower to 'higher' still dominates explanations. The prolific wildlife that has adapted to London's inauspicious habitat at great speed suggests the inter-penetration of human intelligence into that of non-human creatures.

39 Jeremy Narby's book, *The Cosmic Serpent,* describes how the shamans of the Peruvian Amazon tribes take hallucinogenic drugs to communicate directly with snakes (and plants) by which they come to know the properties of plants like curare, tobacco, strychnine, as well as drugs derived from countless others in the most bio-diverse area of the globe. This will be returned to in much greater detail later. His belief, through anthropological studies of many tribes, is that the 'snake' throughout the history of medicine from Aesculapius onwards (and in the symbolic art of most ancient civilisations) is the metaphor for the coiled DNA molecule, the source of all knowledge. This theory seeks to explain the universality of such visions and such understanding, and to account for it through the evolution of consciousness as the mechanism of 'Internal Selection' rather than 'Natural Selection'

See also 'Be ye therefore wise as serpents, and harmless as doves' (Matthew 10:16)

40 The parable of the wise and foolish virgins, the latter missing the coming of the groom while seeking to buy oil for lamps they had failed to make ready.
(Matthew 25:10)

41 The fovea is the point on the retina where vision is most acute. The falcon's eye has two foveae to permit the acute vision required by the extreme parallax it must correct at the extreme height it requires for both spotting and accurately diving upon prey, using velocity to stun its victim.

42 A reference to Ingmar Bergman's film, *The Seventh Seal,* in which death plays a game of chess on the beach with his intended victim, a knight, to delay the hour.

43 Archimedes was asked to assay a crown given to King Heiron 11 which was suspected not to be pure gold, but adulterated with silver. Obviously it could not be melted down. On lowering himself into his bath and observing the displacement of the water, he realised that the density of the crown could be compared by an equal weight of gold: if it displaced less water it would not be as dense, and therefore adulterated. Allegedly this caused Archimedes to shout 'Eureka—I have found it' and all such unexpected insights into the relationships between things have come to be called the 'Eureka' moments. They usually herald a breakthrough in understanding, often giving rise to whole new disciplines in science. Later, these episodes and their chronology will be explained as important evidence for involution.

44 It was believed that if a coin left teeth marks after being bitten it was not gold but lead, a much softer metal.

45 Rodin's *The Thinker* and Michelangelo's *Tomb of the Medici*.

46 String theory is a recent addition to the physics of the invisible. It suggests that there are many more dimensions than those we can perceive, currently estimated somewhere between ten and eleven. It is thought to explain the nature of the universe in which there is a shortage of matter. According to string theory, there are coiled up universes of dark matter that sustain the theories of conservation of energy. It is involution's 'case' that this symmetrical conservation is the inevitable construct of the collective scientific intellect, separated from consciousness. As such it is a reflection of mind: not matter.

47 The cosmic 'soup' is an ill-judged metaphor. At the start of the universe there was no oxygen and no water. It required the evolution of anaerobic micro-cellular organisms to produce oxygen and later water.

48 Neutrons carry no charge and are therefore comparatively inactive.

49 The conservation of energy (and the balancing of equations) needs symmetry on both sides of the = 'the gate'.

50 Infinity or singularities defy the laws of mathematics and the

conservation of energy. Yet the appearances of infinities to balance equations are a recurrent problem, and many mathematical devices are invented to get round them in order to sustain the existing laws of physics. Paul Dirac was guided more by the elegance of mathematical alternatives in deriving his quantum theories than any experimental observations, and was always suspicious of the need for infinities. Again, and in another sense, it would seem that the dominance of mind prevails in the theories about the start of the universe and the infinitely small. Is the collective consciousness creating?

51   The Large Hadron Collider at CERN in Switzerland is currently seeking the 'god particle', the Higgs boson, postulated to exist by the English physicist Peter Higgs. This is the most recent quark conceived to exist and, if found, to answer all questions about creation of the universe. Through its asymmetric influence matter comes into existence, after which the laws of conservation will prevail. At the start of this recent experiment the magnets which are used to accelerate particles failed through some simple structural 'spanner in the works'. (Perhaps caused by a widespread terror that scientists might not know what they were doing?)

52   Babushkas. The Russian wooden dolls of decreasing size stacked one within another, as quarks are envisaged 'stacked' within larger atomic particles.

53   According to the laws of Newtonian physics, entropy is the inexorable tendency for any physical system to 'deteriorate' into maximum uniform disorder, like gas molecules distributed evenly in a container. Quanta were first named by Max Planck as the 'jumps' in the atom from one state to another, in discrete irreducible 'packets of energy'. It was either in one state or another, but never in-between. This was later explained by the orbits of the electron in fixed 'shells' around the atomic nucleus.

By contrast with unorganised matter, evolution took a negentropic path, 'against the grain' of physical systems, with more complex, more organised forms of life evolving. This is now attributed to the stability of the DNA molecule which 'conserves' the improvement which accelerates through time. The *'narrower crags'* describes the ascent of the mountain towards the convergence of man, the single species, in which each major 'jump' towards greater complexity took less time.

54   The '*hare*' in the poet's mind was 'internal selection', the impulse of an organism to behave. Only as a result of spontaneous behaviour (the product of the organism's consciousness) could natural selection act…the '*tortoise*' following after.

55   The subatomic particle's 'charge' accounts for its behaviour, as it does in atomic interactions, but these led to the formation of more complex molecules at the start of time; such as the oxygen molecule combining with hydrogen to form water.

56   Kekulé, the father of organic chemistry, allegedly while sitting dozing by the fire, saw whirling molecules join hands in a ring. This gave him the stable benzene ring—another 'eureka' moment. The Greek Ouroboros symbol of a snake biting its own tail recurs in many forms, the yin yang of the Tao, the anaconda that surrounds the cosmos, and sometimes the split snakes of the caduceus, the symbol of healing. Jeremy Narby asserts that all of these are the symbolic references to the coiled structure of DNA. The evidence is persuasive. (See The Cosmic Serpent. Note 39)

57   Ambrose suggests that it would take a minimum of five genes to change to bring about the smallest change to an organic structure, and that macro changes would need an 'intense input of new information'. The Archaeopteryx, that intermediary dinosaur, half reptile and half bird, has had a lot of argument riding on its wings, yet no intermediary between it and a proper bird has ever been found, nor between it and a reptile. Yet all is not lost if one accepts that neither Darwinism, nor random purposeless mutation is enough to explain why the over-riding climb of evolution has been to produce upward evolution, complex from simple.

Conway Morris suggests that once there is a high degree of complexity, relatively small genetic changes could trigger large morphological ones. A piece of grit can seize an engine or blow the tyre of Concorde. The complexity itself introduces new and accelerated forces for change, and could provide the 'input of new information'. The same pigments can provide very different pictures when mixed in different proportions; the same computer programme can be used by a genius or a simpleton, and DNA is an intricate digital programme, continually being updated. The generations of computers that have evolved in the last thirty years would seem a fruitful analogy about the re-integration caused by complexity alone.

Recent work on the FOXP2 gene which appears to account for human language ability would support Conway Morris dramatically. Although the FOXP2 gene exists in all mammals and birds, humans have a small difference in theirs which manufactures a protein differing in only three amino acids from that of mice. Two of these differences have been shown to first appear in the period when Homo sapiens diverged from chimpanzees, about 200,000 years ago. These small changes act like micro-switches turning 'up' 61 other genes, and 'down' 55 others; these other genes in turn are related to brain, mouth, muscle, vocal cords and breathing control necessary to speech and language. Most recent research is now locating these 'switches' controlling genes in the 'junk' areas of DNA. Suggestion made later about the origins of life would give historic coherence to this phenomenal major jump to Homo sapiens with his languages. Yet birds can imitate sounds; (Parrots not only imitate but understand words) and dogs comprehend meaning; the mechanisms lie 'in potentia' across widely distributed species.

This must be the conclusion from the greatly accelerated rise to complex creatures from a simple cell containing a single molecule. In that sense there are no 'primitive' forms. All the simpler predecessors of complex organisms co-exist (apart from the extinct and over-specialised) and all continue to perform within complex creatures.

58   A central concept in Involution is the collective's determination of acceptable hypotheses which is why genius is always before its time; ahead of the collective. Not until experimentation validates the truth of hypotheses do they enter the collective. Genius is always individual. In this sense Darwin does not qualify as a genius, but a diligent observer. He merely repeated what had been observed by Greek theorists and intermittently thereafter. He just happened to express the concept of the 'survival of the fittest' (which he read in Malthus) in a Victorian age when competition and 'progress' were manifest everywhere, in industrial strife and creativity, in the encounters with so called 'primitive' societies. It was an ethic ready to understand both the interconnectedness and the evident 'superiority' after colonial

expansion encountered 'primitive' peoples. Peter Medawar gives deeper understanding to this conservative tendency of collective science by showing that the pretence of objectivity or deduction is spurious. *A priori* assumptions govern all scientific experiments and observations, determining what are 'worthy' to be observed and thereby skewing what is actually perceived. The convention of reportage in the passive voice only adds to this pretence. The recent gauntlet work by Sheldrake, *The Science Delusion*, seemingly critical of science's hypocrisy in this, unwittingly continues it by suggesting that psi phenomena should be experimentally and statistically investigated. If the great wealth of corroborated intuitive or mystical experiences has not penetrated science for two hundred years, why perpetuate the belief that they are accessible or meaningfully interpreted statistically? It merely invites more of what he criticises.

59   Here lies the nub of the argument for the structure of mind, through the processes of Involution. The making of tools, as opposed to their use (which has always been considered to distinguish primates from the lower orders of animals) resulted in the increasing mastery over the environment, with nature 'tamed' to man's purpose. As a result, the perception of the environment, the 'it' outside of mind severed mind from consciousness, resulting in the two-fold mind of man; the conscious mind of accumulated experience (both collective and individual) from the unconscious mind of the evolutionary experience, stored and recorded in DNA.

60   The evolution of the brain, the most complex structure which endows man with his dominion has taken, in evolutionary time scales, (using a clock face as the entirety) about three minutes of the clock to the midnight now. It remains the scientific belief that mind is the product of the brain. The oldest parts of the brain, the so-called reptilian areas in the caudal region closest to the spinal cord (where through the ventricles the cerebrospinal fluid circulates) connects the higher brain functions with the bodily control of movement, and sensation. The nerves connecting the brain to muscles and senses arise in this brainstem area. They are concerned with co-ordination of muscles, special senses and the automatic visceral and somatic functions. It is as though, through the remnants of the earliest kinds of brain, the later higher functions are moderated, inhibited or excited. The reptilian

brain controls the functions for which no conscious thought is required, and over which it has no control (except in advanced yogis), such as blood pressure, salivation, respiration and smell.

The latest part of the brain to develop, the forebrain with its cortex, is that involved in conscious thought. Although certain areas are identified with some senses and functions, it appears to be much more plastic in re-allocating functions when parts have been damaged. Damage to the earlier reptilian brain, which have widespread consequences, are seldom able to be remedied or 're-assigned'.

Could it be that the wise serpent, or the coiled kundalini of the Vedantas, is the residue of our experience of evolution, which, summoned from the coccyx and rising through the conscious mind (as yogis seek through various meditative practices), is the conscious recovery of evolutionary memory? If so, the serpent would lie closest in our kinship with everything else. This would endorse the claims of the Amazonian Indians to reach wisdom of the inter-relatedness of all life by contacting their inner serpent.

61   T.S. Eliot. 'The Love Song of Alfred J. Prufrock.'

I grow old… I grow old…
I shall wear the bottoms of my trousers rolled
Shall I part my hair behind? Do I dare to eat a peach?
I shall wear white flannel trousers, and walk upon the beach
I have heard the mermaids singing, each to each
I do not think that they will sing to me.

62   John Keats. 'Ode to a Nightingale'

63   It is said that Pythagoras, on hearing the ringing tones emanating from the blacksmith's forge, realised that the pitches of sounds bore a mathematical ratio to one another. He tested this by constructing a monochord; a single-string instrument which, when stopped by a finger at the half way point and plucked, produced an octave higher than the string itself. Halving the 'half' produced a fourth, and other subdivisions produced the third and fifth. These simple numerical ratios are the foundations of Western musical harmony and the intervals are known to string players as the natural harmonics: they can be elicited by merely touching the string at these precise points

and either plucking or bowing. Natural harmonics give rise to artificial harmonics on another set of more complex fractional relationships.

The relationships between mathematics and music was extended later in such concepts as Kepler's music of the spheres and his rejection of the elliptical orbits of the planets, which seemed (at the time) to defy perfect relationship with numbers, although conic sections were later shown to have more elegant mathematical precision. The relationship between sound and speed (in which the pitch of sound alters as it accelerates away— as in an ambulance) later contributed to the speed of light and its red shift as a measure of acceleration in an expanding universe.

Unlike Dawkins, I believe that 'bad poetry' can contribute significantly to the perception of new connections!

64 Dawkins describes genes in such terms. Arms races are the competitions between species whereby evolutionary improvement in one produces improvement in another. The cheetah's increased speed requires the gazelle to match it. Because genes are each 'selfish' in the pursuit of their own survival, and unchecked would counterbalance one another within the single organism, other genes evolve to 'moderate' the effects of powerful genes and 'outlaws' remain inactive until a change brings them into functional purpose. He acknowledges there are 'selfish co-operators as well; 'A species gene pool tends to coalesce into a gang of mutually compatible partners'– Always terms of aggression; even co-operators are 'gangs'.
In contrast, Lynn Margulis writes:-

'Next, the view of evolution as chronic bloody competition among individuals and species, a popular distortion of Darwin's notion of the 'survival of the fittest', dissolves before a new view of continual co-operation, strong interaction and mutual dependence among life forms. Life did not take over the globe by combat, but by networking. Life forms multiplied and complexified by co-opting others, not just by killing them'.

65 Lynn Margulis first suggested that the mitochondria in the animal cell which is responsible for its energy production (and has its own DNA) had been retained from bacteria and co-opted to do what the more complex cell needed doing. Even cells do not re-invent the

wheel! The plant's equivalent structures, the chloroplasts, do the same for the cells of plants, and from the same bacterial origins.

Fred Hoyle suggested that the quantum jumps in sudden species arising, without any intermediaries shown in the geological record, were due to new bacterial DNA arriving from outer space. These, he suggested, could have nudged the sudden increases in complexity by doing business with the DNA forms on earth. A similar idea, 'Panspermia' was put forward by Francis Crick. Hoyle believed such a hypothesis would explain why widely differing species (in evolutionary time as well as structural complexity) had so much in common, such as eyes using the chemical retinol to trigger the visual nerve impulses, in insects, in the octopus and in vertebrates. It would explain why plant substances have the ability to affect the chemistry of animals. Morphine and quinine are examples which a single cosmic source might have introduced at an early enough time for its chemical influence to be retained. Genetic engineering uses a similar process routinely. Viruses change the way the infected host cell works to make more viruses, and in response the hosts evolve antigens to repel them. This would explain the species-specific nature of disease.

In 1957 the interstellar dust was shown to be remarkably similar to clouds of bacteria and obey the same refractive indices as do dried bacteria. Moreover bacteria are shown to survive the kinds of temperatures they would meet on entering the earth's atmosphere, provided they were small enough. The eggs of insects would meet the smallness criterion. The ability of micro-organisms to survive extreme radiation would imply that exposure to such radiation had previously happened, as was shown by exposing Micrococcus radiophilus to more x rays than would kill a human, only to observe its ability to repair the damage to its DNA. How would this capacity have evolved if life was purely earth based without any need for it? Meteors have been shown to contain bacteria very similar to life forms on earth.

66   The 'codons' of the DNA molecule are combinations of the four bases, Adenine, Guanine, Thymine and Cytosine, which combine to provide 64 variants, each coding for specific proteins. This four syllable alphabet speaks the words of protein language and all of creation. Much more about DNA later; in detail in Footnotes to Canto the Ninth

Footnotes to Canto 2

67 An 'abstract' at the start of a scientific paper gives the outlines of the argument to be put forth, some idea of the range of evidence to be presented and anticipates the conclusions to be drawn.

68 'Pom falugo perflasmatosh'. The sound of a stone tossed into a pond and the rising and dispersal of water following. It was quoted to the poet as a child as an illustration of Greek onomatopoeia, the ability of words to capture the sounds of what is described.

69 Evolution is traditionally thought to be divergent, with the proliferate tree of life arising from a single stem (cosmic soup!) The dominant single species, Homo sapiens, if embodying all the stages that preceded him, and with a mind connected to all, could equally make evolution convergent; more a cone than a tree.

70 The acceleration of evolution's changes remains perplexing, since the rate of mutation in genes is roughly stable. Most confer no advantage. The most complex organs (including the human brain) develop more rapidly than much simpler organ systems. To suggest that thought (through action) influences the changes in genetic structure removes the improbability of acceleration and random change being a sufficient explanation. Instead it gives action spontaneous and momentary 'purpose', and increasingly discriminate action quicker purposeful consequences.

As things stand science does not admit of any possibility whereby behaviour can influence organic structure. Jeremy Narby has suggested that under hallucinogenic drugs shamans take consciousness down to molecular levels to gain information related to DNA. DNA emits photons (which are electromagnetic waves) and because they have their origin in biological molecules they are referred to as bio-photons. All cells emit such bio-photons at an almost constant rate (100 units per second) and DNA appears to be the source of this emission. Interestingly, these emissions have a wavelength corresponding to the narrow band of visible light. Narby suggests that these emissions are likely to be responsible for the visions produced by hallucinogens, as well as the visions seen by shamans of talking snakes and plants.

Bio-photon research generally conceives of these emissions as being cellular communications (which might account for the collective plankton or slime mould behaviour). Narby quotes Fritz Albert Popp's opinion that consciousness could be 'the electromagnetic field constituted by the sum of these low intensity emissions'. (This would be the consciousness 'field' equivalent of the background radiation that gave the necessary support to the big bang origin of the universe.) Although low in intensity, they are at constant frequency from all DNA containing cells, which means everything in the biosphere. Could this be the nature of the Akashic Field? (Alternative and very recent hypotheses can be found in Footnotes 375 and 376.)

Quartz crystal (widely used in shamanism and found in pre-historic sites) is known to greatly amplify these bio-photon emissions due to the stable structure and frequency at which it vibrates.

DNA is also a crystal, but an irregular or 'aperiodic' one except in the junk sequences which align similar shaped components one above another. These areas of DNA would vibrate as quartz does, due to the regularity of the periodic structure at these points.

This could explain both the emission of bio-photos and the 'reception' or 'reading' of those emissions…in other words a mechanism whereby cells communicate with one another and explain consciousness.

Narby supposes that this could explain the source of shamanic knowledge, derived from the biosphere to which DNA crystal is sensitive. In short it provides an explanation of the retention of 'junk' DNA, for its receptivity to all about it.

Within Involution's hypothesis it would explain the mechanism whereby cumulative memory is stored and able to be read under hallucinogenic drugs, or more generally in dreams and deep reflections, or within the self-forgetful state of contemplation, or love. (see Carl Jung)

71 The double helix of DNA in the nucleus of all eukaryotic cells is a mirror image chain of base pairs linked together by hydrogen

Involution

72   Keats's poem 'The Eve of St Agnes' is quoted more than once in this Canto.

> St. Agnes' Eve— Ah, bitter chill it was!
> The owl, for all his feathers, was a-cold;
> The hare limp'd trembling through the frozen grass,
> And silent was the flock in woolly fold:
> Numb were the Beadsman's fingers, while he told
> His rosary, and while his frosted breath,
> Like pious incense from a censer old,
> Seem'd taking flight for heaven, without a death,
> Past the sweet Virgin's picture, while his prayer he saith.

73   '*A posteriori*' (from the latter) or. '*Post hoc, ergo propter hoc*', literally 'after this, therefore because of this'. Science's insistence that it takes its conclusions from the evidence fails to fully acknowledge that the design of experiments is an act, a priori, that determines what is likely to be found. See also Peter Medawar's opinion on the conventions of pretence to objectivity.

74   While evolution describes the transformations of matter from the start of time to now, involution describes the transformations of mind moving inexorably back towards the beginning. We are now collectively returned to the fields of thought mastered by the Greeks, the nature of mind, the origins of matter, and the causal relationships between them. This was referred to in the introduction as supporting Whitehead's understanding that the direction of mental causation moves from the future towards the present and is always 'now', the physical creation that results will always be 'ago'. This phenomenon, including the critical role of thought experiments in science (prior to measurement and evidence) will be more fully annotated through later Cantos.

75   '*Newly flexed*' because it uses the same evidence to draw different conclusions.

76   T.S Eliot. 'The Love Song of Alfred J. Prufrock.'

'Oh, do not ask, 'What is it?
Let us go and make our visit

In the room the women come and go
Talking of Michelangelo'

77   Nerve conductivity is electro-chemical in nature.

78   The eukaryotic cell, the fundamental building block of all organic life, is considered the least explicable 'jump' in the evolutionary process. (See Lynn Margulis) Once achieved, however, it's protected 'integrity' of concentrating the factory of life, with its processes and sub-processes, makes all else possible.

79   Mitochondria are the components, believed bacterial in origin (and later enclosed within the cell membrane) in animal cells that provide the energy, storing oxygen in the red blood cell haemoglobin. Their equivalents in plant cells are the chloroplasts which transform sunlight into chlorophyll.

80   *'Radioed'* will be understood better after the footnotes to Canto the Ninth. Sound waves seem to be inherent in the way DNA interacts, and seemingly retains the records of interaction. (See Garjajev)

81   Improvisation. A spontaneous musical response in which keys and rhythm, although provided, do not constrain the melodic or ornamental possibilities. A raag is an Indian 'form' of improvised music in which the time of day, the numbers and natures of the instruments, determine what keys and scales express the appropriate mood. Some are morning raags, some evening, some played after midnight. Why a raag expresses evolution is because the ascending scale permits certain notes and only certain intervals, and the returning scale different ones, and the whole expresses something 'beyond'. The ambition of each player is to prolong the return to the initial silence by stretching musical survival in a kind of drawn out 'tease'. The orgasm is endlessly delayed, but the suspense towards it grows through the interactions between the players.

82   *'Solo and accompanist'* acknowledges that each creature is both

Involution

selfish in the pursuit of its own survival, but equally provides the means and environment which fosters or drowns out another's performance.

83   The slime-mould inhabits forest soil, feeding on bacteria and yeast which they locate by gradients of folic acid. Shortage of food causes them to aggregate in vast clusters to form a differentiated single multicellular organism which, from a ' body', sends out a 'stalk' on which is clustered a bolus of spores. This differentiation is into a genuine organism (rather like a slug) to safeguard the survival of the single spores. From the analysis of specific proteins it is thought that the slime mould diverged after the plant-animal split: it retains greater complexity than yeast but simpler than plants and animals.

Unlike the gametes (sex cells) of plants and animals, the slime mould has its genetic material in one strand only (a haploid) rather than the diploid (duplicated) characteristic of later cells. It therefore represents a kind of pre-sexual amoebic stage in which reproduction requires a collaboration of many to form an organism, rather than the specialised development of part. It seems an Ur organism which retains capabilities lost in later complex creatures. Complex organisms retain this capacity only in the nucleus of mitosis-dividing cells retaining the potential for specialisation through cell cultures, hence the name 'stem cell'.

A recent discovery in the University of Texas (reported in 'Nature'), Dictyostelium discoideum feeds on bacteria which it carries from place to place and 'cultivates' by seeding it to reproduce in times of scarcity.

84   This emphasis on the capacities of the slime mould is simply to illustrate that not only orientation towards food, but 'sensory' awareness, and self protection can be achieved by chemicals and responses to chemical gradients alone. It does not stretch the point unduly to propose that the structure of DNA might be altered by similar experiences in more complex creatures. Penrose has now described a microtubule lattice structure that could, by digital computation, provide 'awareness' in single non-nucleated cells. This capacity in each cell is later 'orchestrated' to afford sympathetic coherent awareness in later multi-cellular organisms. (See Footnotes 375 and 376.)

85   The four commonest elements in the tail of comets are the same as

Footnotes to Canto 2

those on Earth: hydrogen, carbon, nitrogen and oxygen. They reflect in their relative numbers the same proportions as in the biosphere. Bacteria-like forms (Pedomicrobium) were found by Hans Pflug in the centre of the Murchison meteorite. Fred Hoyle claims that these are the 'fossilized and shrivelled remains of life outside our planet'.

Hoyle goes further to suggest that Earth has periodically and constantly been bombarded by life-potential forms from outer space. The rain of meteorites known to have struck the Earth 65 million years ago (and to which the relatively swift extinction of the dinosaurs is commonly attributed) led to the emergence of the mammals and the rapid proliferation of the major Phyla in relatively short geological time. That impact is recorded in a band of a rare element, Iridium, not found elsewhere. We already know about the decimation of human populations (Amazonian Indians, or the !Kung) when introduced to Europeans against whose diseases there was no immunity.

Hoyle proposes that the possibility of intermittent genes, added by space-forms to earth-developed domestic ones, would inject new potential and accelerate the sudden jumps observed in evolution, accounting for rapid advances followed by much slower incremental change.

One of the major objections to classical Darwinism is that the geological record does not provide the intermediaries we should expect, if slow incremental improvement was all there was to distinguish species. Punctuated equilibria (jumps followed by relative stasis), a theory proposed by Eldredge and Gould, would be explained by intermittent new genes 'infecting' the stabilized gene pool, and to which they would not have built up defences.

86   Francis Crick used the term 'Panspermia' to echo Hoyle's hypothesis of life originating in space.

87   The Archaeopteryx, half reptile, half bird, was once thought to be the definitive 'missing link'. It was believed time would produce the intermediaries from reptile to Archaeopteryx and from the latter to birds, but those intermediaries have never been found. The argument against graduated change is further supported by the fossils of ancestral flies millions of years old but almost unchanged in their modern descendents.

Insects, of all Phyla, would have had the best chance of transfer from space (as eggs or gene-bearing gametes) being small enough to withstand the pressures of entering the earth's atmosphere at high temperatures. Perhaps those genes had already stabilised, and the creatures bearing them were already finely adapted, thereby explaining the lack of change.

Bacteria have been found in radiation chambers, in volcanoes and in every conceivable extreme temperature on earth. The adaptation by the eukaryotic cell of bacterial structures (Mitochondria and Chloroplast) pays tribute to the heritage of micro-organisms in space and still fundamental to the processes of cellular life throughout the living kingdom.

88   Jeremy Narby's book, *The Cosmic Serpent,* deals with the reliance of the Amazonian Indian people on medicinal plants to cure disease. The correspondence between diseases to which we on Earth are prone, and simultaneously the naturally occurring chemicals to cure them, is another argument used by Fred Hoyle for the evidence of space origins. He argues that chlorophyll is not the most efficient converter of the sun's spectrum (being unable to harness the green spectrum which as a result is reflected back). To be truly efficient across the whole of the light spectrum leaves should be black.

However, chlorophyll has building blocks in common with something in interstellar dust that absorbs the sun's rays similarly. So, having arrived, that precursor was simply available to do the job required, rather than evolving to do it better.

89   Viruses inject cells and interrupt the cells' own programme. This infectivity spreads rapidly. Infection results in antibody formation, so terrestrial plants and animals are effective detectors of new disease. The intermittent epidemics of diseases for which no antibody defences are present do suggest new invasions; affecting shepherds in remote areas and Bedouin out of contact, just as much as urban concentrations. Studies have shown that rural populations have fewer defences against diseases because the antibodies have not previously needed to be developed, due to infrequent encounters between them and sources of infection.

90   Carbon containing molecules can exist in mirrored forms. Amino acids are among them but, for an unexplained reason, naturally

occurring amino acids exist only in one form, the L form, (although both L and D forms can be equally easily produced in the laboratory). Strangely, amino acids arriving in meteorites also show this L form dominance. There has been no explanation for this. Any asymmetry in the structure of matter raises questions.

91 Richard Dawkins played God playing evolution on a computer (in order to demonstrate the processes of small incremental changes leading to large alterations in phenotypic effects). From a limited number of 'genes' and successive generations between which small mutations of adding and subtracting were inbuilt, the original 'forms' through (his) artificial selection, required 100 mutational steps between an inefficient catapult 'Y' shape, and the succession of 'biomorphs' that resembled every kind of fly or beetle. This was to demonstrate the effects of cumulative selection *in itself* as adequate explanation for adaptation. Of course it speeded up mutation a hundred fold and kept the environment (the computer programme) both fed and watered which guaranteed survival of all the progeny, of which he selected the ones to breed. This of course leaves out all the factors (slow rate of mutation, changing environment, finding your own mate, competing with others) which would prevail naturally.

92 The juice of the King Coconut is identical to blood plasma. This and other examples (the treatment of many diseases by Penicillin, the responses of the human nervous system to Morphine, the use of Quinine to treat malaria) are, in Hoyle's opinion, inexplicable in terms of the Darwinian hypothesis. The genes of the curative substance and that of the disease susceptible to it would have evolved widely separated in time. However, if new genes were introduced from space that affected separate lines evolving in tandem (and thereafter continuing to do so) this congruity between plants and animals would be explicable.

93 Taxes are the simplest responses to external stimuli. Most plants are positively phototaxic. Chemo taxes are the responses to chemicals (positive and negative) by which simple animals find food and avoid threats. Balancing such stimuli differentially on either side of the body enables complex orientation, as bees show to the angle of the sun. Bats respond to sonic frequencies and echo location to locate prey

and avoid objects at high speeds. Complex behaviour patterns (such as the courtship dances of birds and fish) are achieved by a succession of sequential responses to stimuli which lead ultimately to mating. The gradual interiorising of the sequences (originally requiring the external stimulus) leads increasingly to autonomous behaviour and, in the higher animals, exploration and problem solving.

Black cab is a pun on the London taxi and the *knowledge* is the name given to that required by taxi-drivers who have to pass a rigorous test on the London streets and areas before being licensed to carry passengers.

94   Schütz (1949) showed that young storks on their first migration showed an ability to maintain a constant orientation (SSE). Perdeck in 1956 demonstrated that young starlings needed previous exposure in the company of others to return to where they had previously wintered. Learning in the latter was needed to underpin the migratory pattern, whereas in the storks the entire sequence had been 'interiorised' and is now controlled by genetic factors. These examples support the hypothesis that behaviour (with its cognitive components) is progressively 'involved'. The increasing autonomy of behaviour and the methods of learning as one ascends the evolutionary ladder is the most cogent evidence for involution in sub-human species.

95   Dawkins. R. *The Extended Phenotype*.

96   Lamarck (1744-1829) proposed, in his law of 'use and disuse' that the 'Inheritance of Acquired Characters' accounted for the slow incremental changes, by which he believed the gradual improved adaptation in evolution occurred. This was in direct conflict with Darwinism which saw random mutation as the sole method of evolutionary change, conserved by natural selection. Lamarck was much lauded in Soviet Russia and its programmes of social engineering in which these principles were applied to effect a predictable improvement. Agricultural and metal workers would become stronger and more adapted to their assigned roles, scientists would spawn better scientists. The system would interlock all the hierarchies required by the State in a perfect machine.

Due to this association with the ruthless application of his beliefs, Lamarck is now wholly discredited. The 'Theory of Involution'

Footnotes to Canto 2

approaches a Lamarckian hypothesis inasmuch as it suggests a two way interaction between the experiences encountered by a creature and the storage of that experience in DNA, leading to more and more independence from the stimulus-response interactions characteristic of simple taxic reactions. The role of mind as being increasingly important in determining advance runs counter to the fundamental tenets of Darwinism and evokes a Lamarckian phobia, except in scientists unafraid of opprobrium like Laszlo and Sheldrake who propose the Akashic field and morphic resonance to explain the influence of the past upon development and acceleration.

Where involution differs is that its 'interiorisation' is subconscious, not necessarily immediately manifest, and not necessarily conferring an advantage to the individual. Being retained in memory everywhere, it contributes to the substrate consciousness that leads to collective 'ascendency'. Therefore it is the 'all' in 'one', and the 'one' in all.

97   Of the genes in the human genome, only 2-5% is believed to determine the development of the individual. The remaining 95% is called 'junk', thought simply to retain the evolutionary history through which man has evolved. Junk shows repeated sequences or strings of the four bases that make up the alphabet of DNA, but interestingly enable the determination of close relationships that are used in criminal identification. The 5% that are known to code for development through protein manufacture are held in common with all others, the junk sequences are held in common with those closest in familial relationship.

It could be that the similar tastes of non-identical twins, raised independently, is coded in the 'junk' DNA. Junk is given greater prominence later in this hypothesis (and succeeding Cantos) as the likely storage and influence of mind generally. Instances of twin telepathy, and simultaneous symptoms of injury appearing in a non-injured twin are now becoming widely known and have been well documented (and experimentally tested) by G.L. Playfair. It would appear that close related relationships establish almost a common 'mind' which directly and non-locally affects the body. This 'connectedness' established by shared experience, and its subsequent mutual resonance is what is proposed to lie throughout the biosphere.

Involution

98   String theory is the most recent development in physics and attempts to provide an explanation for the irreconcilable gulf between relativity and quantum theory. There is not enough matter in the universe to explain the conservation of energy. String theory proposes that in tandem with the universe of matter of which we are aware (with three dimensions and time providing the fourth), there are many parallel universes with ten or eleven dimensions which are undetectable to our senses, because our senses are limited to the universe we evolved in. These parallel universes make up the required deficiency.

String theory has been necessitated by the exhaustion of our mental constructs, and the success of our penetration of the universe outside ourselves. We have yet to investigate the universe of consciousness!

99   Michelangelo's bound slaves in the Accademia in Florence show the partial emergence of the slaves in relief, still imprisoned in the sculptor's block.

100   In preparation for fertilisation the gametes (sex cells) divide and retain the single string of DNA (as opposed to the double chain in every other cell) so that on fusion, as a result of fertilisation, the full complement is restored. It is this meiotic division that enables the recombination of genes and the variation on which evolution depends. Sex linked genes are those on the single Y chromosome in males (or on the X which may be carried by either men or women) which hardly ever get recombined, because they are permanently held in union with sex chromosomes, of which the male has only one copy (one X and one Y). The so called 'Royal Disease' of Haemophilia, which passed through the royalty of Europe (from Queen Victoria's offspring) stemmed from a single X linked gene which was not counteracted (by a second X chromosome in the males) and therefore manifested in the males as the disease, although the females carried it unobserved. The second X chromosome in the females mitigated the effects of the deficient recessive one. It was therefore the sons who were sickly.

101   'Ontogeny recapitulates Phylogeny' is an evolutionary maxim. The development of the individual (ontogeny) goes through, in embryo, the stages of the evolutionary path (phylogeny). Embryology has proved the most fertile area of study that illuminates the gradual

evolution from one form to another.

What is here emphasized is the same parallel in the stages of behaviour. Gesell (1954) drew attention to this. 'In all vertebrate creatures the general direction of behaviour organisation is from head to foot. This sequence in motor patterning in the human infant clearly shows this law of developmental direction. The lips lead, the eyes follow then neck, shoulders and arms, trunk, legs, and finally feet. Reflexive motor patterns give way to learnt patterns, first of the more simple type such as the imitative behaviour of the one year old, the trial and error problem solving of two, the insightful behaviour of three….development entails continuous interweaving of patterns and component patterns. The organism is forever doing new things, but learning to do them in an old way, reincorporating at a higher level what it has approximated at a lower one. The structure of mind is built up by a kind of spiral cross stitching.'

Perhaps the word 'spiral' was fortuitous in 1954, emerging just as the ultimate spiral was being unravelled. That DNA spiral is what connects the nested complexity, or holarchy of evolution.

102   '*Needlessly*' in terms of genetic information. Only the sex cells determine the evolutionary path. As mentioned above, the slime mould gets together for asexual reproduction, each spore having only a single copy of the DNA required. It begs many questions as to why in the evolution of sexual reproduction (given the specialisation of other organs to take care of specific structures like liver, kidneys, lungs etc) that every single cell needs to retain the genetic recipe. Dawkins mentions that it might have been more efficient for the organism to limit reproduction to only certain cells, with an over-all command centre. The different development of parts from each other implies that each requires all. Leaves require stalks and each develops in harmony and differently. Sheldrake suggests this need for different information requires his hypothesis of morphic resonance; so that a would-be 'leaf' develops differently from a would-be 'stalk' (since the instructions are identical in both). Involution suggests that the additional distinctive information could lie within the 'junk' sequences. Sheldrake does not seem to suggest any additional role for DNA beyond protein synthesis, perhaps because he puts undue emphasis on the fewer genes present in man as compared with simpler organisms. If complex discriminatory behaviour has been 'interiorised' making mankind

able to spontaneously react (due to his complex nervous system and its memory) would he not need fewer 'instructions' than simpler creatures and therefore fewer determinant genes?

Every cell has a memory: every cell has the information to re-create the whole. Cells are 'fractals' of the hologram that is the man. One of the limitations in postulating a different causality is the now embedded belief that time runs only from past to present. However, if fertilisation brings with it the future individual, then 'activating' parts of the DNA not concerned with protein manufacture but with other faculties, mind, intelligence, or artistic gifts, would 'moderate' or influence the development of the resultant embryo. DNA could be the interface, whereby what Sheldrake calls 'morphic fields' shape the development of the particular organism. The determinant 'matrix' influencing development would however be pre-biological, or mental, or post biological and memory. The irradiation of potato stalks by DNA modulated lasers, which then produced tubers (See Garjajev's work in footnotes to Canto the ninth) now suggests that DNA contains all that is needed to 'instruct' the differential development of different parts of an organism, in the right order, at the right time.

103   Although Athens in the golden age was a hobbledehoy town of small houses and warren streets, the central square, the Agora, was wide and shaded with plane trees.

Democritus (460-370BC) with Leucippus first proposed that matter consisted of atoms (impossible to 'cut') that were the 'primary corpuscles' and irreducible to smaller components. They attempted to reconcile the observed fluidity and changes in states, (generation and corruption) and the constancy of things that were not altered by observation (unlike secondary qualities that were a matter of opinion).

This was in direct opposition to Heraclitus's (500BC) ideas of everything in perpetual flux of 'unity in opposites'. Heraclitus's central idea was that all parts were connected with every other. 'Harmonia' is omnipresent.

According to Heraclitus there is a parallelism between the identity of the mind and that of the reality it grasps. The meaning of the world can be grasped by 'introspection'. Human reason has the power to know the language of creation precisely because its own operations are conducted in the same language. Behaviour and structure of the world and the soul run parallel. The role of language in DNA is developed

Footnotes to Canto 2

in the footnotes to Canto 9. Involution as a theory could be said to be Heraclitus restated in evolutionary dress.

104   Praxiteles (350BC) and Phidias (c 430BC) were sculptors, the latter believed to have both supervised and contributed to the Parthenon friezes.

105   Bakelite: An early precursor to the plastics first created by Leo Baekeland, a dark brown brittle material widely used for early valve-radio casings: The sort of radio that sat on 1930's mantelpieces and was used for the crackling transmissions during the 1939-45 war.

106   The axons are the long neuronal 'cables' surrounded by myelin sheaths which insulate the relatively long extensions of the neural cell and increase the conductivity. Connections between nerve cell axons are made through the branching dendrites of adjacent cells at the synapses. The nerve impulse crosses these connections in one direction and is the result of changing polarization due to excitation. Whether the excitation is sufficient for the travelling impulse to cross the synaptic 'gap' depends on many factors, the distribution of many synapses stimulated at once, the presence of inhibitory impulses etc. Drugs that alter consciousness act largely on synaptic resistance or lower the synaptic thresholds across which impulses travel.

107   What is proposed here is that brain cells (having all the information common to others) may be set to transmit impulses by a sympathetic faculty, just as a string on a violin will vibrate to certain natural harmonics set up elsewhere. Iain McGilchrist's comprehensive analysis of the influence of brain asymmetry on the 'eras' of human thought suggests that the left hemisphere, which follows linear, categorizing, language dependent rational methods, seeks power of narrow clarity between either-or (as in science). This, he suggests, has dominated certain periods of history (post Socratic Greece, Rome, the Reformation, the Enlightenment and now to a destructive degree Modernism), reducing art to minimalism, dislocation, context free, and value free meaninglessness. In contrast, the right hemisphere comprehends the wholes, the context, the inherent alternatives, and is happy with apparent contradictions, the 'both-ands' of understanding exemplified in pre-Socratic Greece, the Renaissance, and the Romantic

Involution

Periods.

It could be that each hemisphere has developed to respond to complementary aspects in which the individual has been pitted and had to survive in the context of all. McGilchrist's appeal, in presenting compelling evidence, is for us to recognise the disproportionate role of the left hemisphere since the Industrial Revolution. Power and control through bureaucratic institutions, has afforded materialism, insecurity and greed at the expense of everything that makes life worth living; art, community, relationship. This hemispheric asymmetry is most pronounced in the capitalist West dominated by fear and the need for power. The pattern of integration whereby the left hemisphere, having analysed the parts, restores them to the right for deeper understanding (raising it to a higher level… 'aufheben') has been lost: once the left succeeds in wresting power there is no corrective or negative feedback that causes it to relinquish control. The Master of the right hemisphere has been usurped by his Emissary, the left.

108 Olivier Messiaen claimed to 'see' musical sounds in colours and shapes. He represented birdsong widely in his compositions. In his Turangalila Symphony he made extensive use of an instrument called the Ondes Martenot, a kind of hybrid organ (capable of swelling the sound from piano to forte) vibrating on a single note (as does the string player using vibrato). This keyboard instrument takes great skill to play and seems almost to fill a pre-existent gap in the symphony orchestra's repertoire of sound. The blending of the senses (synaesthesia) is characteristic of infant perception: the right hemisphere dominates for four years before the left hemisphere, through the acquisition of language skill, becomes ascendant. The examples of artists throughout this book are likely to be examples of right hemisphere dominance. The immersion into consciousness by creative artists may well be mediated by the right rather than the left hemisphere of the brain, though this is my conjecture. The 'translation' of that immersion will require the mediation by the left hemisphere's assessment of the prevailing market, now so narrow, cynical and controlling there is little room for formal art of any kind.

109 Pablo Picasso

110   The modern computer works entirely on programmes built from two digits, 0 and 1, hence 'digital'. Dawkins likens the gene's storage to a digital computer, more a 'recipe' for a cake than a 'blueprint' for a building. The cake, having been baked, cannot revert to its ingredients, unlike a building.

Involution applies this analogy even more tightly. The 'outside' and the 'inside' of consciousness (material reality and mental reality) are mirrors separated by the division of the mind through time.

Footnotes to Canto the Third

111   Olfaction- the sense of smell.

112   'Anthropomorphic' refers to attributing human qualities to non human creatures or inanimate things. Like a clock 'grinning' which soon follows in the poetic text. 'Rückblick ist jetzt streng verboten'- Hindsight is prohibited.

113   Radio-active decay is the rate at which certain elements lose radio activity. The half life of any radio active element is the time taken for half of the initial quantity to decay. This is a constant, so it takes as long for the initial amount to be halved, as the next (original) quarter, the next (original) eighth etc.

114   Zeno of Elea (c470BC) argued (in his paradoxes of motion) that a runner given a length to traverse must first run half of it, then the half of the half and so on. From this it could be deduced that he would never reach the end. This argument applied to the 'Achilles and the tortoise' paradox whereby Achilles, giving the tortoise a head start would never catch up.

115   Einstein's thought experiment, in which he sat on a beam travelling at the speed of light and saw the clock slowing and stopping, allegedly led to the special theory of relativity.

116   This refers to the supposition that impulsive action by the individual was the determinant of the evolutionary process.

117   David Bohm wrote of the 'implicate order'. It is now believed that only 5% of DNA is 'used' in determining the development of an individual. (More recent analysis limits it to 2%). Involution suggests that the remaining 95% of 'junk' DNA contains the future

development of man, because it retains all of the past. (It seems more than co-incidence that it is now thought that only 5% of brain activity reflects 'mind', self aware mind). The remainder does what? Keeps the individual connected to the world of (sub) consciousness? By the end of this epic, through re- tracing the chronology, what is revealed is science's incremental recollection of that 'junk' memory, to the apex of the cone where mind and matter are united. In short, thus far, evolution has been clearing the bin. Now is the start of a new cycle which is perhaps what underpins all the 2012 Mayan Calendar, prophesies and apprehension. Deep down, was this anticipated? The apprehension of the future inbuilt in consciousness?

118   Recent theories of physics suggest that the laws of physics are contradicted by the apparent insufficiency of matter in the universe. This has led to the Mverse hypothesis (dominantly by Stephen Hawking… mysterious? multiple? mistaken?) coiled up within, and imperceptible to us, because our minds are governed by the one of which we are part. All the other 'verses' provide the necessary requirements to maintain parity or symmetry throughout, in other words to sustain the existent framework of an independent material universe. String theory, by describing 'smooth' transitions, eliminates the angles of change in which infinity makes its appearance, and therefore persists in keeping an infinite universe at bay. If neither 'string theory', nor the Higgs particle stand up more than metaphorically the 'standard model' will have to be abandoned, and a separate material universe replaced with 'a universe of mind stuff' as suggested by Arthur Eddington. Many look forward to it.

119   There are early Victorian pictures of would-be balloonists seated astride a horse. The relevance of this is to the limitations of the habitual which govern the possible. The balloon could not lift the horse, as the material world now cannot 'lift' the imagination beyond a construction governed by the material. All the hypotheses offered to overcome the paucity of matter are material hypotheses, despite the unknown nature of creation.

120   This roughly translates as 'Always something new out of Africa'

121   *Ur-citrus*. The German prefix 'ur' suggests original, archaic.

Involution

122   Olduvai Gorge in Tanzania (the Grand Canyon of prehistory on the Serengeti) is where Louis and Mary Leakey worked and discovered skeletal remains of Homo afarensis. Oldowan is a term now accepted in palaeontology for the methods and findings established there. Taung is the cave in which the fossil recognised as Australopithecus by Professor Raymond Dart was discovered. Sterkfontein and Swartkrans were other sites in which fossil remains confirmed Dart's bi-pedal proto- human discovery.

123   Lucy was the name given to a partial skeleton found at Hadar in the Ethiopian Rift valley, of a small female dated between 4 and 3 million years ago, which, from the knee joint, indicated that bi-pedalism originated much earlier than previously thought. From this and similar finds, Africa was named as the cradle of mankind, where the great apes were transformed by slow degrees into Homo sapiens, through the erect stance and the consequent freedom of the hands.

124   Turkana Boy was a skeleton, almost complete, found (by Kamoya Kimeu with the help of the Leakeys in 1984) on the shores of Lake Turkana in northern Kenya. His dates were 1.8 million years ago. He was an immature adult with much more human proportions, arms no longer than ours; he stood 1.62 m tall, but with a much smaller brain case. 'Ergaster' meaning 'working' was the name reserved exclusively for the African species of which Turkana Boy was the most complete example.

125   Caledonian crows enhance their diet by using twigs to dig out larvae from trees. A Bonobo has been taught to 'read' and signal from symbols, similar to cuneiform words, on a vast card, both concrete objects and conceptual ideas. He was able to point to sequences that conveyed 'I want-to-go-in-the-car'; 'I-need-a-hug'…so more advanced than a three year old human child; what underlay this ability was not merely interpreting the given but conveying his desires through a choice, somewhere between reading and writing.

126   Khiam points. The earliest type of Neolithic projectile named after el Khiam in the Judean hills of Palestine. They are distinguished by a notch on either side of the triangulate base, which made them suitable later as spear heads.

Hemudu is thought to be the oldest site for the cultivation of rice by 5000BC. It lies in the lower valley of the Hangzhou River in the Yangtze valley. Spades have been found fashioned from the shoulder blades of cattle. In addition knives, pestles, fishing sinkers and spears, as well as shuttles, needles and spindles indicated weaving.

Sophisticated adzes have been found in Polynesia for the working of bark cloth. Most are made of basalt

Obsidian is relatively rare, but its hardness made it very valuable and traded. It is present in some quantity in Turkey but it is found as far afield as Jordan. This suggests the exchange of materials suitable for tool manufacture was widespread.

127   Recent research by Sheffield University has suggested that tranquil environments, occasioned by water, boost brain performance.

'Tells' is the name given to early settlements in Southwest Asia in which huts sit above a mound created by centuries of domestic debris. Skulls found in this debris suggest that early societies buried their dead in the floors of their houses.

Pictogram derives from the rough approximation of a 'picture' to represent a thing. Cuneiform meaning 'wedge-shaped' refers to the tool that created the script of Bronze and early Iron Age writing (circa 2400BC). Cuneiform seals and tablets were found in numbers at Uruk, Lower Mesopotamia (now Iraq).

Tell Halaf in Northern Syria gave its name to widespread characteristics of Southwest Asian culture from 6000 to 5400BC

128   To enlarge upon Leonardo's Vitruvian man, here is his analysis of the relationship between the human figure and geometrical proportions and forms: 'From the chin to the starting of the hair is a tenth part of a figure. From the chin to the top of the head is an eighth. From the chin to the nostrils is a third part of the face. If you set your legs so far apart as to take a fourteenth part from your height, and open and raise your arms until you touch the crown of the head with the middle fingers, you must know that that the centre of the circle formed by the extremities of the outstretched limbs will be the navel, and the space between the legs will be an equilateral triangle. The span of a man's outstretched arms is equal to his height.' (Leonardo's notebooks. Quoted by Arthur Koestler *The Act of Creation*, p 332) The sense of the mathematical microcosm in man (which involution extends into

his microscopic DNA) has had a long and reputable history.

129   Analemma refers to the figure of eight that the sun transcribes over a year, caused by the tilt in the earth's axis.

130   Teilhard de Chardin drew emphatic attention to the quality most distinguishing man from the rest of the created world, as 'knowing that he knew'. Other animals 'know' (through instincts or other neurological patterns) but do not 'know that they know'. This is normally taken to mean that man's use of tools, and the corroboration allowed by language are the means whereby man comes to 'know that he knows'. Involution gives a different dimension to this hypothesis by suggesting that the recovery of memory establishes that man 'always knew', but subconsciously, until the recovery becomes conscious.

131   Kundalini is the name given to the axis of consciousness running from the earthly 'base' of the anus, up the body to emerge (when consciousness is released from the physical) through the 'thousand petalled lotus' on the crown of the head. The seven 'chakras' of ascending consciousness are threaded up the kundalini, each associated with a different level of consciousness. The *daisy chain of hydrogen* links this kundalini with the hydrogen atoms that link the paired bases of the DNA molecule. The critical importance of the structure *between* the mirrored strands of DNA is further elaborated in the footnotes to Canto 9.

132   The great reconciliation between relativity and quantum theory is what has been sought for the past decades. Both provide accurate accounts of their respective worlds of the macroscopic and microscopic but are yet to be reconciled in a single overall Theory of Everything.

133   The search for the 'god particle' predicted to exist by the English physicist Peter Higgs has prompted the research being done in Switzerland in the Large Hadron Collider. If found, it is believed it will provide a link between relativity and quantum theory and between the non material and the creation of matter. The Omega minus particle predicted (and found) by Murray Gell-Mann, (through a process he called 'the eightfold way') was, in the 1960s, believed to hold the same potential solution. Gell-Mann coined many of the terms by which

quarks (his invention) were distinguished; 'strangeness' and quantum 'colours'.

134   The Rape of the Sabines is a subject much painted. Rape in this context meant 'kidnap' and it referred to the Roman army taking the Sabine women captive as hostages to bring about the conquest of the Sabines, a very war-like people.

   Space and time were intimately bound together in Einstein's theories of relativity, to become space-time, an entirely different concept created by density of matter and its curvature of space, in which time varied according to both speed and distance relative to other things. Electricity and magnetism were similarly spliced in electro-magnetic field theory (further detailed in Canto the eighth).

135   The biblical 'talent' was a coin. It is also an ability, a gift, and both senses are intended.

136   A quotation from Keats's sonnet 'On First Looking into Chapman's Homer' which merits quoting in full for its relevance to this verse.

   Much have I travelled in the realms of gold, / And many goodly states and kingdoms seen; / Round many western islands have I been/ Which bards in fealty to Apollo hold. /Oft of one wide expanse had I been told / That deep-browed Homer ruled as his demesne: / Yet never did I breathe its pure serene / Till I heard Chapman speak out loud and bold: / Then felt I like some watcher of the skies / When a new planet swims into his ken; / Or like stout Cortez, when with eagle eyes / He stared at the Pacific—and all his men / Look'd at each other with a wild surmise— / Silent, upon a peak in Darien.

Footnotes to Canto the Fourth

137   Rubicon. The name of the river crossed by Julius Caesar which marked the boundary between the Northern Roman province and Southern Italy. The crossing was the declaration of an act of war. It is now a metaphor for crossing a point of no return.

138   The tripod, as well as being a three legged stool, also refers to the three legged altar at the shrine of Apollo at Delphi on which the priestess sat to deliver the oracles.

139   The Bodleian Library at Oxford is one of the oldest libraries in Europe, holding a copy of every book published in Britain. It took over from the older Duke Humfreys Library (14th Century) in 1598.

140   The first known astrolabe is attributed to Hipparchus (150BC). It is an astronomical instrument that enables the calculation of spherical motion in astronomy. It was greatly advanced by Muslim scholars later.

141   Tom Tower, designed by Christopher Wren, is the bell tower affording entrance to the great Tom Quad of Christ Church College, Oxford. Its bell always rings a fraction after all the other bells in Oxford.

142   A Quotation from Shelley's 'Ozymandias' 'I met a traveller from an antique land / Who said: 'Two vast and trunkless legs of stone / Stand in the desert…' It ends with the lines 'My name is Ozymandias, king of kings: / Look on my works ye mighty and despair!' / Nothing beside remains. Round the decay / Of that colossal wreck, boundless and bare / The lone and level sands stretch far away'

143   Johannes Kepler, having tried to reconcile his astronomical

observations of planetary movement with the harmony of the spheres, was appalled to find that the movement traced an elliptical path. Before Newton, he did not have the benefit of gravitation to explain this, and described his discovery as a 'cartload of dung'.

144   A reference to Rene Descartes' 'Cogito ergo sum' (I think, therefore I am). This does not necessarily imply that one exists because one thinks, but rather that thinking is all one can be certain of, the starting point of examination. Descartes' scepticism restored primacy to thought as the source of knowledge and is generally considered to be the founder of the scientific revolution, as well as an outstanding mathematician and philosopher in the early development of rationalism, which contributed greatly to science's eradication of intuitive process, at least officially.

145   *Pilots without legs*. Douglas Bader, a fighter pilot, who, although losing both legs, continued to fly raids over France in WW2. The Red Baron was the nickname of the First World War German pilot, Manfred Albrecht von Richthoven, who as an aristocratic 'Freiherr' was nicknamed both for his title 'von' but also because he painted his aircraft red. The Wright Brothers, Orville and Wilbur, are credited with being the first to develop powered fixed wing aircraft, although this is widely contested.

146   Seleucid refers to the last three centuries BC in Mesopotamia with complex systems of theoretical astronomy, knowledge of the zodiac, and a list of eclipses used later by Ptolemy. They were considered much more advanced than the later Egyptians, whose application of geometry to the measurement of fields and time (the shadow clock) was primarily practical. The Egyptologist, Schwaller de Lubicz, contests this long held orthodoxy through his deciphering of the Temple at Luxor in which the building and its hieroglyphs betray a sophisticated mathematical understanding of the triangle and the 'golden mean', pi and phi. He suggests that Pythagoras acquired his understanding from the Egyptian priesthood, as did Moses, and all Gnostic sects since. (See 152 below)

147   Ionians usually refers to the Asiatic Greeks from Asia Minor (delimited as lying between Ephesus, in what is now modern Turkey, to

Involution

Halicarnassus). The Dorians were further south still. They preceded the Athenians (as far as records are reliable) in their cultural speculations. Thales (624-565BC) was part Phoenician (roughly the Canaanite areas on the coast of the Mediterranean, the Fertile Crescent) and a knowledgeable sailor applying geometrical calculations to navigation and distances.

148   Anaximenes (c570BC) from Miletus followed Anaximander and Hecateus (also of Miletus). He travelled widely throughout Europe and Persia, drawing a map (500BC) of North Africa, Europe and the Middle East to the Caspian Sea.

149   Heraclitus of Ephesus (540-475BC) was the great philosopher, whose all embracing concepts were widely influential, although difficult to comprehend. Essentially enantiodromia described the tendency of everything to become, through perpetual flux, its opposite. All matter was in a perpetual state of flux and a union of itself and its opposite. None of his writing survives except through his influence on others. In a way Heraclitus' idea underpins the hypothesis of anti-matter and dark matter that prevails today.

    Lavoisier (AD1743-94) resolved the arguments on conservation by conducting very careful experiments and showing that boiling water (changing it from one form to another as steam) would, after condensation, and with the addition of any solid particles remaining, restore the initial quantity of water. He showed similar conservation of matter through combustion and burning and discovered and named Oxygen, realising that water was a combination of oxygen and hydrogen. This eliminated phlogiston theory and set the course of most scientific experiments thereafter.

150   Democritus (470-400BC) was a contemporary of Socrates (who rejected everything he thought). He was the first to propose the atomic theory of matter, with atoms (irreducible solid matter) separated by void. Atoms were identical, and matter differed only in their arrangement. Democritus' theory largely underpinned the Epicurean view of temporary and random creation.

    Dalton (1766-1844) examining the expansion of gases (all equal with an equal increase in temperature) derived his atomic theory to explain this. He calculated that although atoms were too small to weigh, their relative weights could be compared by knowing the constitution of a compound, and comparing those with one another.

151   Herodotus (484-425 BC), a Persian by birth, travelled widely and wrote his famous *The Histories* (it meant 'enquiry') describing different peoples' nature and habits. He is often called the father of both anthropology and history.

152   Pythagoras of Croton (582 BC) was an Ionian from Samos who settled at Croton in Southern Italy. He founded an ascetic brotherhood that held the view that numbers existed independently of mind and outside time. Mathematics (which meant simply 'learning') had a central role in all things. Heaven was a numerical musical scale and expressed the correspondence between the working of the mind and of nature. Ten was the perfect number and the sphere the perfect shape, which governed speculation of the planets and planetary movement until Kepler. The pentacle with its ten mirror points and identical angles was of special significance to the Pythagoreans, later considered 'debased' through its 'magical' abuses by astrologers.

According to the Egyptologist, Schwaller de Lubicz, Pythagoras had learned his hermetic understanding of the universality of number in Egypt, where a secret tradition of priesthood encrypted their understanding in the Temple of Luxor. Not only in the hieroglyphs inscribed on the walls, but the relationships of these to the joints in the stones where they were placed made the temple itself the model of man; the innermost sanctuary being the 'head'; the outer vestibules representing the parts of the body; in all, the 'Anthropocosmos'. The 'temple' was man himself. Certainly the secrecy of the Pythagorean Brotherhood would support such a connection, and its influence upon later adherents such as Kepler was a persistent theme in the search for numerical and harmonic relationships throughout the natural world. The hermetic tradition persisted under many guises up to and including Newton and the Invisible College that preceded the Royal Society. Hermetic practice has always been seen as a threat to the authority of any institution, religious or scientific, because it endows the individual with the knowledge that precludes any need for them.

153   Arnold Schoenberg (Vienna 1874-1951). His early compositions

Involution

were Wagnerian in idiom with his monumental Gurre-Lieder Symphony the apogee of this period of composition. Later he based his work on the dodecuple scale which used all twelve tones of the chromatic scale on equal terms and with equal emphasis (unlike the octave scale which had simply chromatic embellishments). Most Western Music used the octave scales with their 'relative' scales modulating composition. Other cultures use other scales which gives the distinctive 'voices' to national music such as the pentatonic (5 tone and characteristic of much early and folk music) or chromatic (12 tone, often attributed to the Pythagoreans who tuned the intervals slightly differently, and early Chinese music used it as a basis for tuning.)

154   Philolaus (480-400BC) was a Pythagorean and one of the first writers about their beliefs. Plato in his Timaeus refers to him and he influenced Copernicus in the 15th Century who re-discovered his heliocentric hypothesis.

Nicholas Copernicus (AD1473-1543) studied at several Italian universities and became familiar with Pythagorean ideas. He studied astronomy mostly from the observations of others. He wrote *de Revolutionibus* which reconciled the contradictions by positing it was the earth that circled the sun, but it received little recognition at the time, and was not published until three hundred years after his death.

155   Parmenides (Born c.515BC) has been the subject of much recent study, as perhaps the first embodiment of the philosopher-king in its original religious sense. Peter Kingsley has written copiously on the neglect of the true meaning and contribution of Parmenides and these notes are derived from his, admittedly unorthodox, but persuasive re-examination of Parmenides' fragmentary original sources. Influenced by Pythagoras, Parmenides was a Phocaean, from Phocaea on the Turkish coast. Phocaeans fled to Velia in Southern Italy, where Parmenides lived, and later to Massalia (Marseilles). Phocaeans were closely bound to one another and distinguished by the tradition of Iomantris, healer-prophets who continued the Pythagorean hermetic practices of deep meditation, or incubation, through which mystical union with the Goddess gave access to the underworld of Apollo, and Justice and Wisdom, in order to directly convey the nature of reality to mankind. They did not seek to escape the natural world of the

senses but to perceive it clearly as illusory, but necessary, in order that mankind should exhaust its potential to realise his true eternal nature.

Parmenides' famous poem in three sections describes travelling with the 'Daughters of the Sun' through whose persuasion Justice at the gates of the Underworld admits him into the presence of the Goddess. The words from the Goddess describe the fork in the two roads offered to man; one which leads to her and truth, and one that keeps him confused and lost in the world of illusion, or matter. The very precise instructions given have both incantatory and instructional components, designed to strip away illusion, until she restores the necessary deception of the world. The world of the senses reveals a seamless eternity providing man follows it 'all the way through all there is'. 'Dying before you die' was the intention to gain access to the experience that death was illusory and consciousness beyond intellectual understanding. In that sense, his life was dedicated to the practical enactment of divine relationship with Apollo and Being/Creation.

The world and thought were bound as one, and therefore Parmenides' mysticism was not one that repudiated but incorporated the world and thought as a single phenomenon. 'To think something is to make it exist… Whatever you can think about has to exist, if it did not, you could have no thought about it'. The faculty most needed to comprehend this apparent duality is mêtis, a cunning and comprehensive awareness of everything at once, and the trickery required to both 'read' the deception and respond appropriately. (The kind of animal awareness leading to spontaneity suggested as the basis for Involution.)

Kingsley suggests that Plato deliberately distorted the teaching of Parmenides in order to discredit Zeno, his adopted 'son' (i.e. his appointed initiated successor), to make Plato by implication his heir, and to replace the practices of initiation with intellectual hypotheses of which Plato was to prove both master and dictator for fifteen hundred years. From the plane of Parmenides' revelation Plato built his plane of rational argument. From the point of the hypothesis of Involution the resurrection of Parmenides' true message now is exactly what one would expect, the return to the man ultimately uniting heaven and earth, at the start of recorded human history, before the later Messiahs fulfilled similar intercessions and instruction.

Involution

156   Empedocles (490-430 BC) was born in Sicily a little after Parmenides, although obviously aware of his writings as well as the practices of the Pythagoreans. What remains of his poetically expressed philosophy also suggests a claim to the immortality of a God and a capacity for prophesy and healing. His cosmology dealt in greater detail with the nature of illusory and deceptive creation whose cycles were also the fate of the Soul. This work was addressed to Pausanias, his young disciple (the beloved initiate), indicating that mankind is powerless to discover his nature without guidance from a teacher (as in many mystical Gnostic schools). He defines the natural world as composed of the four 'roots' (later called 'elements' by Plato) which start as separate, divine and perfect, but through the deceptive allure of Aphrodite (Love) are made to attract and blend with one another creating incarnation (blood) to entrap the Soul. The 'fall' is twice (as in the biblical succession; first the fall of Lucifer and his angels followed by the Expulsion of Man); from immortal separation to mortality and then further separated through sex and beguiling union.

These unions of matter are torn apart by Strife which is the counterpart to Love and which sets the Soul free to re-ascend the aether of Union. Just as Plato is alleged to have distorted Parmenides, so too does Kingsley suggest that Aristotle was instrumental in diminishing the transcendental philosophy of Empedocles and laying greater stress on his rhetoric than on his meaning. Empedocles seemed (to generations of later modifying opinion) not to have meant to give Strife such a creative importance in the Soul's liberation, yet this is to distort the needful opposition to Love which imprisoned the soul in matter. The 'Creation' remains created by love but a prison from the deeper reality all the same.

157   According to Kingsley, the Phocaeans undertook the 'imitation of Hercules' in pursuing prolonged journeys or rigorous disciplines to reach the wisdom available through self-negation. (See Pytheas, Note 183 below). Even without this narrower meaning, the 'Herculean' nature of these first thinkers merits the appellation.

158   The Breakfast at *'First Light'* is meant as counterpart to the Last Supper. The same intention to save mankind from illusion and helplessness by the early 'disciples' scattered about the Aegean and Mediterranean.

159   Xifias is a Swordfish, a common Aegean speciality.

160   The 'Fighting Temeraire' was a term accorded a gunship that fought valiantly at Trafalgar, and also the subject of a beautiful painting by Turner in which the ghost of the galleon is being towed by a tug to her destruction. The *Laughing Temeraire* is a similar tribute given here to the ship of consciousness, leaving the breadth and ease of the relationship between God and man as manifest in Pre-Socratic thought, into the darker passage of divided intellectual debate to follow. The Return (of consciousness) from the Exile (of the intellect) will take the exhaustion of the journey here recounted.

161   Hippocrates (460-370BC) of Cos came from a family of physicians. He transformed the traditions of scientific procedure. He was said to have lived to 100 and known to have been learned, observant, humane, with a reverence for his patients as a source of information. The Hippocratic method of inductive reasoning, of patient observation rather than mystical hypotheses, held that generalisations could only emerge from repeated experience. The contemporary Religion of Science, in which a constant chain of cause and effect is supposed fundamental to explanation, stems from the Hippocratic School, further developed and entrenched by Aristotle.

162   Aristophanes the dramatist with others of his contemporaries such as Aeschylus, Sophocles and Euripides donated all the terms used in drama to this day; comedy, tragedy, dialogue, character, metre, and chorus. Cicero said 'In learning and in every branch of literature the Greeks are our masters'. Brevity and simplicity and an economy of line were instinctive in Greek literature, possibly because they had no prior examples to guide or to adapt.

163   Hippocrates of Chios (430BC) wrote the *Elements of Geometry* (the same title as Euclid's great 13 book treatise). He discovered that a semicircular lune on the edge of a circle bounded by an arc of 90 degrees is equal to the area of the triangle formed by the same chord. This insight enabled the calculation of the area of a circle.

164   The Academy was the famous school in Athens that formed around Plato.

Involution

165   The Oxbridge colleges (which include both Oxford and Cambridge) were patterned on monastic buildings with central courtyards, often cloistered, and sleeping quarters for the scholars off stairwells. All have chapels.

166   The Peripatetic School was that formed by Aristotle, succeeding his original Lyceum. Peripatetic and pedant merely referred to the ambulatory habit of discourse from master to pupil. Cloisters continued to provide for it in monastic buildings. Later pedant gained the pejorative meaning of narrow, tram-lined and merely derivative of prior opinion.

167   Pallas Athene was the patron Goddess of Athens, whose temple was the Parthenon.

168   Socrates (469-399BC), whose words and thoughts were recorded by Xenophon, emphasised above all the predominance of soul, and its survival of death. Plato's *Phaedo* was a tribute to his master, Socrates, who also appears caricatured in Aristophanes' *The Clouds*. In his *Apology* Plato gives an account of the Delphic Oracle's answer to a question put to her by Chaerephon who asked whether anybody wiser than Socrates existed. The Oracle answered that no-one was wiser. According to Plato, Socrates set out to prove the Oracle wrong by holding analytical dialogues with as many as possible. He could only discover those who believed themselves wise but who were not. He concluded that the only reason he was wisest was because he knew he was not. Socrates took part in a battle in Potidaea, after which he seemed overcome by a trance and stood immobile for twenty four hours. It was after this 'transport' that he devoted his life to converting others to care for the health of their souls. As nothing survives of Socrates himself, all records are subject to distortions particularly Plato's, his pupil. His evangelism following his 'conversion' echoes the search and practice of Parmenides, Empedocles and the Pythagoreans, and he marks the transition between the experiential and contemplative path to truth and the intellectual, post Socratic division that followed, which we trace through the remainder of this work.

169   The Agora was the open air market place in Athens (or in any city).

Footnotes to Canto the Fourth

170 Two sword lengths…the safe distance between two sides of an argument. Also a reference to the distance between the opposing benches in the British House of Commons.

171 Aristotle was born at Stagira, near Mount Athos in Asia Minor. It was common to call notable men by their place of birth, and Aristotle was commonly called the Stagirite.

172 Plotinus, a Neo-Platonist, will make a later entry, a mystic who revived Platonism in 3rd century AD in Rome.

173 Just as Peter Kingsley has spent years re-evaluating the contributions of Parmenides and Empedocles (suggesting that both have been distorted first by Plato and then by subsequent classicists, devaluing their contribution) so too has Ken Wilber done much the same for Plato whose alleged gulf between Ideas and Forms led to the dualism between illusion and reality. Instead, Wilber claims that Plato's concept of God, unlike the 'ascendant only' God of Aristotle was aware of God's 'descendant' role in matter, and therefore His non-dualism. This was perpetuated by Plotinus's understanding of Plato and through Plotinus the influence on Aquinas and contemplatives ever since. This current, new absorption in the Greek philosophers is cogent evidence of Involution's return to the origins of Western thought, and their renewed relevance (whether one is sympathetic to the reappraisals or not).

174 Euclid (330-260BC) also wrote a later *Elements of Geometry*. He was called to the Alexandrian Academy by Ptolemy, ruler of Egypt. Trained by Plato, his thirteen books of 'The Elements' was the fundamental text for all students for twenty two centuries. It concerned much more than geometry; drawing on the works of Plato and Hippocrates of Chios it covered conic sections (after Apollonius), theory of numbers (after Pythagoras), irrational numbers, optics, fallacies in reasoning and elements of music.

175 Thomas Aquinas (born in Aquino, Italy AD1225-1274) was the great Dominican theologian who was canonized fifty years after his death. His concept of an unchanging, unlimited, omniscient God stems from Plato, although much of his work was translations and

Involution

modification of Aristotelian ideas about the role of reasoning as the basis of natural theology by which man could come to know those aspects of God not revealed by faith or revelation. The Timaeus is considered the most obscure of Plato's texts.

176   The *Phaedo* was Plato's tribute to Socrates' life and thought. How accurate a portrait he paints of Socrates is now debated.

177   The Schoolmen is a nickname for the schools of scholasticism that developed between AD1200 and 1400 (during the establishment of the early universities) that saw attempts to fuse Arabic and Jewish thought with Latin Christendom. Scholasticism was dominated by the recovery of Aristotle from the Arabs and his concept of the unchanging perfection of the heavens (co-incident with outer space). Dante's hierarchical schemata in his *Divine Comedy* are a kind of literary scholasticism; which was ruled by clerics, Dominicans and Franciscans, who concentrated on the natural philosophy of Aquinas. Among the most famous was Roger Bacon.

178   The Stoics (named after the 'Stoa', the corridors off the Agora where they met and talked) were known for strident attempts to harness all knowledge in service to the idea that natural forces determined everything; everything is determined by everything else. Force acts on matter, matter on force. Deity acts on reason, reason becomes law. Their cosmology linked 'pneuma' (being) to ether which gave rise to the four elements, air, fire, water, earth. Fate determined events which led to the great reliance on astrology. Stoicism became the dominant philosophy of the Roman Empire, consistent with its application of force and law.

179   The Epicureans, named after Epicurus of Samos (342-270BC), took their stance from the atomism of Democritus. Uninterested in broad philosophy, Epicurus concentrated on dispelling religious fears and their inhibiting effect on free lives. Epicureanism spread through both Asia and Rome.

180   Lucretius (95-55BC) was a Roman follower of Epicurus. He wrote *On the Nature of Things* which held that creation was due to the interaction between atoms without intelligence. He was one of the

first to believe in the indestructibility of matter. He held a simplistic view on the ladder of evolution and developed a particular interest in haphazard calamities, lightning, water-spouts, volcanoes and sudden pestilence.

181 Theophrastus (372-287BC) made important botanical studies; gave Greek vernacular names to parts of plants such as 'carpos'(fruit) 'pericarpium' (seed vessel) and he recognised that plants had two 'sexes'.

182 Autolycus (360-300BC) worked on the geometry of the sphere for astronomical purposes.
    Dicaearchus (355-285BC) was interested in physical geography and drew maps. He was the first to introduce latitude from the Pillars of Hercules to the Taurus and the Imaus ranges of the Himalayas.

183 Pytheas (360-290BC) relied on the works of Dicaearchus when he set sail from his native Marseilles round Spain to Cadiz and the Atlantic seaboard to Cornwall and round Britain. From Kent he crossed the channel to Scandinavia, returning to Marseilles. During this journey he fixed other latitudes with great accuracy and observed the reliance of the tides upon the phases of the moon. According to Peter Kingsley, Pytheas went further. He also was a Phocaean who undertook a herculean journey to the Arctic Circle (believed the home of Apollo) and who described the 'solid sea' which was 'neither walk-able nor sailable'. Such a journey, in imitation of Hercules, was, (according to Kingsley), similar to the mystical denials of early Christians 'in Imitation of Christ' and in keeping with Pythagorean hermetic practice. Visiting the North Pole was an extreme act, but it was a visit to the believed seat of Apollo, in Pytheas's case an external journey, the complement to Parmenides' internal one.
    He reached an island off the coast of Scandinavia (called Avalon) on which amber was abundant, the tears of Apollo's 'daughters of the sun'. Pytheas's claims of the influence of the moon on the tides was for centuries disputed, as in the Mediterranean there are no tides, and most Greeks never travelled beyond the Pillars of Hercules. His accurate marking of latitudes enabled later sailors to validate them almost unchanged. Amber, called 'Elektron' becomes very important towards the end of this epic, as the fundamental particle of connection

and complexity. In this account of early Phocaean practice and belief we find the roots of subsequent mythologies Christian, Arthurian, Scandinavian and Scientific. So we return to a unity of mythology as well as of science. I suspect Kingsley, as an original new interpreter, is not universally accepted, against the grain of established views.

Footnotes to Canto the Fifth

184   This reversal refers to the central hypothesis that after the emergence of man, progress is through the gradual recovery of memory; from holistic science of the Greeks, through the fracturing of the sciences to re-approach the holistic theories of relativity and quantum mechanics.

185   After the Roman conquest of Alexandria, the Great Library was burnt, apparently by accident in AD48 when Caesar set fire to ships which spread. Whether this was the Ptolemaic Royal Library or one of the other public libraries is uncertain. Others attribute its destruction to the Emperor Aurelian three hundred years later. The sack of Alexandria by Abd'l Latif of Baghdad in AD642 may have been when certain of the library contents passed to the Arabic speaking world of Persia.

186   Matins, Lauds and Primes, were three of the regular services for monastic institutions throughout the day. The early Celtic Christian settlements on remote Scottish and Irish islands preserved and copied ancient Greek and Latin texts and were largely responsible for their later transfer to Charlemagne's monastic institutions in France c.AD800.

187   'Professor Reductio'- Richard Dawkins has defined 'units of cultural inheritance' (memes) as analogous to genes in their capacity to replicate themselves 'from brain to brain' and sometimes, (like viruses), to occupy a mind against the mind's disposition, like a jingle tune. Anything that replicates by imitation may be called a 'meme' and they 'thrive, multiply and compete within our minds...our minds are invaded by memes just as ancient bacteria invaded our ancestor's cells...' (*Unweaving the Rainbow*)

188   Eleusis was a town northwest of Athens (now part of it) on the

Involution

Saronic Gulf, in which the religious cult of the Eleusinian mysteries took place annually. It centred on the myth of Demeter and Persephone in which initiates were introduced to the evidence of conscious life after death by initiation.

Prospero. The 'magician' in Shakespeare's *Tempest*, the last of his works; like Michelangelo's final pieta, a kind of self portrait. Prospero is served by Ariel 'the tricksy spirit' of the air and Caliban, the unredeemed earthly animal, who only curses in the language he's been taught. Prospero is here invoked as a signal of the intelligence that might lie behind the play of the world. He ends the play with 'We are such stuff/ As dreams are made on, and our little life is rounded with a sleep.'

189   Gat. A slow growing evergreen shrub, chewed as a stimulant throughout countries of the Middle East and Africa, also spelled ghat. The Egyptians believed it to be a divine food which induced hallucinations and put them in touch with the Gods.

190   Lucretius (95-55BC) See Footnote 180 to Canto the Fourth.
       Varro (116-27BC). Originally influenced by Plato, he later turned to Stoicism, writing an encyclopaedia of all the sciences, outlining nine disciplines: grammar, dialectic, rhetoric, geometry, arithmetic, astronomy, music, medicine and architecture. This classificatory approach was typical of the Roman mind. Varro was employed by Caesar to arrange the great stores of Greek and Latin literature. He wrote *On Farming* at eighty, again mostly collected from the writings of others, enlarging on separation and distinctions between things.

191   Pliny (AD23-79) dedicated his *Natural History* to the Emperor Titus. It puts man as central to the natural world; everything else's 'uses' are subject to that anthropocentric doctrine. He relied on Aristotle in his classificatory approach to the varied forms of life he encountered through the strange creatures brought back by the Roman conquests. Pliny met his death at the eruption of Vesuvius by deciding to leave his ship, anchored in the Bay of Naples, in order to observe the eruption and the destruction of Pompeii at closer hand.

192   The dioptera is an instrument for measuring angles, estimating heights and relative distances between distant points. A graduated

Footnotes to Canto the Fifth

circular table is mounted on a column on which it can be rotated in the horizontal plane, as well as its mounting being rotatable on a vertical plane.

Abacus comes from the Greek word for 'counting table'. Originally grooves in a level table held beads or stones, 'calculi', for making rapid arithmetical calculations. It is still used in markets in Persia and Africa, now more commonly columns of beads that slide on wires within a vertical frame. Herodotus refers to its use in ancient Egypt and it was previously set to the sexigesimal (proceeding by sixes) number system of the Sumerians in 2700BC.

193   Following the advice of the Alexandrian mathematician, Sosigenes, Julius Caesar altered the Roman calendar with its four year cycles of variable length (355 days, followed by 377, by 355, by 378) to the Julian calendar of 365 ¼ days for each year, with a leap year at every fourth. He renamed the month Quinctilius Julius– our July– and Sextilius became his successor Augustus's August. This calendar remained in use until Pope Gregory substituted the Gregorian calendar in 1582.

194   Much more than an inspired plumber, Vitruvius (c10BC) was concerned with public health and the siting and drainage of buildings. The advanced sanitation of Roman buildings, its supply of fresh water and its sewers contributed significantly to its public life, leisure and health. The Roman sewer, the Cloaca Maxima has parts still in use today. Rome was supplied with 300,000,000 galls of water a day through its aqueducts.

195   Cicero (106- 47BC), a Roman administrator, orator, jurist and cultured speaker of both Greek and Latin, introduced the Romans to the Greek Schools of philosophy. His letters are renowned for their elegant prose style and perceptive commentary on Roman personalities. He visited Athens in 79BC where his friend Atticus introduced him to the Greek sites at Rhodes and the Peloponnese, including the tomb of Socrates.

196   The son of Zeus and the Pliade, Maia, Hermes had a variety and changing priority of responsibilities. As well as the 'herald' or messenger between Olympus and mortals, protector of shepherds

and farmers, his mobility, both literal and metaphoric, made him the arbitrator of boundaries between gods and dreams, the soul and Hades (to which he was given free passage) and the tricky world of commerce between men.

197   The Dominican Order (often called the Domini canes– Hounds of God) was founded in 1216 in France by Saint Dominic to respond to the need for a preaching order in the rising cities (rather than the monastic seclusion of the Benedictines). Sometimes called the Black Friars (for their hooded black cloaks) they spread very quickly to England; in Oxford by 1221. They concentrated on doctrinal issues, and incorporated women in separate convents at an early stage. The challenge to Catholic Doctrine by the Cathars or 'Albigensian Heresy' meant the focus on doctrine became central to their role as theologians. Thomas Aquinas was the quintessential Dominican. Later they were very influential as Bishops Cardinals, courtiers in ecclesiastical circles and in the Inquisition, approved by Pope Gregory the 9th which was subsequently authorised by Innocent IV (1252) to carry out torture to prevent heresy and schism.

198   Plotinus (AD204-270) trained in Alexandria in what was the school of Neo-Platonism, taking its ideas back to Rome. He drew on both the ideas of Plato and of Aristotle, with an admixture of Roman stoicism to derive the doctrine of the analogous universes of the microcosm (Man) and the macrocosm (the Universe) in which his Neo-Platonist view was that the universe had been made for man. It therefore resonated easily with church doctrine on the creation and as a result became a serious rival to Christianity. He resurrected the Platonic 'Idea' in the 'Form' which governs matter as its ideal. After his death Porphyry, his pupil, published the *Enneads*, a collection of his very influential thoughts which, for centuries, widely influenced church doctrine and other schools of philosophy.

Although he travelled to India and Persia he was essentially a contemplative who believed creation came about through thought. Bodies are 'phantoms like a reflection in a mirror…the perceptible 'here' is a product of the thinking of the intelligible 'there'. The highest part of the human soul is linked with the Intellect 'there' and will return to it. (In this theory of involution the term 'intellect' is used only for the divided and divisive part of the analytical mind.

The 'return' is to the whole undivided consciousness.) For Plotinus, reincarnation was the result of choices made 'here'; in that sense similar to Indian concepts of karma.

Ken Wilber attributes to Plotinus the significant preservation of Plato's essential sense of Divine unity and the non-duality of both ascension to Spirit and descension of Spirit into the material creation, (both Agape and Eros), and the 'full circle' alluded to in Involution's process and journey. All mystics, having experienced this unity, express it differently but perceive no conflict between the Divine and the Creation. Only the intellect argues the toss, and creates the gulf between science and religion.

199 St Augustine (AD354-430) was Bishop of Hippo Regius. His influence on the Church was due largely to his philosophic analysis and reflections on doctrinal difficulties. He set up a community of disciples in what is now Algeria and wrote on ethics and the problem of evil, but after becoming Bishop all his writings tended to have a pastoral purpose in preventing schism and heresy. He was greatly influenced by Plotinus (see above note 198) 'The true God is the author of things, the illuminator of truth and the giver of happiness'.(City of God 8.5) 'God is not only the creator of things but the means of our knowing of them…knowledge is enlightenment'. Neo-Platonism also governed his views of man's relation to other creatures inasmuch as the connection was through souls. On scepticism he anticipated Descartes ('If I am wrong, I exist') and although he believed man had freewill, he needed God's help in exercising it. Sex was something that brought man low and celibacy, although difficult to maintain, should be aimed at.

200 Kepler's three laws of planetary motion were extended by Newton in his more universal laws of gravitation. What kept the moon in orbit was terrestrial attraction, which explained the elliptical orbit about the earth, (due to a falling towards the earth when closer, and a tangential speed further away) just as a stone falling from a tower does so at a speed inversely proportional to the square of the distance from the centre of the earth. This universal attraction (gravitation) between all masses united earthly and heavenly phenomena within a single 'system'.

Involution

201   Until Einstein there was a lingering belief in some kind of 'universal aether' (ether) in which bodies moved, and through which light waves were propagated. An ingenious experiment by Michelson and Morley in 1887 showed (contrary to expectation) that the earth was 'at rest' in the ether. Light moved at the same speed whether its source was approaching or receding, this would not be the case in an ether 'wind.' Later this conclusion was questioned by exact measurements which re-introduced the possibility of the ether, but by then Einstein's relativity was generally accepted and no amendments to it were acceptable. This 'fashion' for dogma (from whatever source) is central to the control of so called 'objective' science. The constancy of the speed of light was what defined the limits of calculations (and the measurable universe) until very recently. Einstein's special theory of relativity relegated the 'ether' as meaningless, as was any motion or otherwise of the earth within it, or indeed even the earth's position (neither 'here' nor 'there'). Relativity has no absolute referential locus, in space or time. Thus Einstein also floated free from 'either' (Kepler or Newton) and from 'ether'.

202   In Western Asia, Greek was replaced by Syriac and the Nestorian church was established. It was considered heretical and persecuted by the Byzantines, so it emigrated to Mesopotamia then, after the sixth century, to Persia where it established its capital, Gondisapur, known for its dedication to Greek texts, including Plato, Aristotle, Euclid, Archimedes, Ptolemy, Galen and Hippocrates. These were all translated into Syriac.

Byzantium in the 7th century was invaded by the Arabs, and Gondisapur became the new Islamic centre of learning. With the rise of the caliphs, Islamic learning absorbed the Greek sources through the translations by a family of Nestorian scholars from an extended family, the Bukht-Yisheu, over seven generations, who translated the Syriac into Arabic. From 750 – 850 Arabic began to dominate the translation of the ancient Greek texts. A school and a library were established by the seventh Caliph Al-Mamun in Baghdad specifically to concentrate on translation and accumulation, particularly of the medical sources from Galen and Hippocrates.

203   Jabir (AD760-815), called Gerber by the Latins, was concerned with occult relationships between numbers. He was a Sufi who is thought

to have known Greek and, as well as his alchemical transmutation interests, made genuine chemical discoveries described in his *Book of Properties*, which classified minerals into Spirits: sulphur, arsenic, mercury, camphor and ammonium and Metals (7). A laboratory of Jabir was discovered, not long ago, on the Tigris.

Rhazes (AD865-925) is considered the greatest Arabic writer on alchemical matters, as well as what were later new chemical discoveries. Although he believed in the transmutation of metals, he defined the necessities for a laboratory, as well as being the first to define substances as animal, vegetable and mineral. He closely observed the interactions between chemicals, classifying them accordingly. '*Transmutations poor retort*' uses 'retort' both in its laboratory sense, the stand on which chemical reactions occur, and as the 'answer' given by aeons of alchemists to the hope of turning base metals into gold.

Avicenna (Ali-ibn-Sina, AD980-1037) denied the possibility of transmutation. He wrote *The Canon of Medicine* which was studied as a medical text until modern times; far in advance of anything available to western medicine. All three men made medical contributions on disease and systematic classification of cures and symptoms.

204 Al-Khwarizmi, a Persian, used the Arabic numerical notation where the value of the number depended on its position; i.e. the difference between 0.1 and 1.0. This was originally Indian in origin. He introduced the word 'algebra' which in Arabic means 'restoration'.

205 Alhazen (AD965-1038) of Basra developed optics and wrote a treatise, *The Treasury of Optics*, which altered the prevalent belief that the eye sent out signals in the form of visual 'rays' to the object viewed. He suggested instead that the form of the viewed object that passes into the eye, is 'transmuted' (i.e. refracted) by the 'transparent body' (i.e. the lens) to create vision. His calculations for the propagation of light and colours, illusions and reflections, including the calculation of the angles of incidence, refraction and reflection, which at the time involved complex mathematics using the hyperbola, came close to the idea of magnifying lenses not developed until Francis Bacon's work in the 13th century.

206 Maimonides (AD1135-1204) was born in Spain but, exiled from Cordoba, spent most of his life in Cairo. He was a Jewish philosopher,

Involution

a jurist and physician. A critic of Galen, he wrote extensively on hygienic matters, as well as writing Arabic commentaries on Jewish Law, and Hebrew translations. In his *Guide for the Perplexed* he reveals that the anthropomorphic approach conceals God's absolute perfection. Biblical creation is more probable an explanation for the divine multiplicity of differences evolving out of simplicity. It demands that man should pursue his likeness to God by perfecting himself, living in peace with one another and other creatures. He believed Islam and Christianity have spread monotheism through the world in preparing for a Messianic age (seemingly derived from the Israelite prophets).

207   Printing probably began in the Far East, but the moveable-type press of Johann Gutenberg in Mainz (c 1447) spread rapidly. It first required the printing of the Greek classics before it made much impact on science, although theology and philosophy benefitted more rapidly as did etchings and art. Albrecht Dürer was one of the first to exploit its commercial possibilities. *Hoi Polloi;* Gk. meaning 'the many' but generally indicating 'the rabble'.

208   Constantine the African (1017-1087) was a Tunisian, who arrived in Salerno in about 1070 where he became secretary to the Norman conqueror. He later retired to a monastery where he spent his last years translating Arabic works, including Arabic Greek sources into Latin. With him, Alphanus (Archbishop of Salerno) resurrected many ancient Greek texts, which continued to be consulted even after better translations were available from the scholastics such as Adelard of Bath and Gerard of Cremona who translated Ptolemy and Avicenna. Aristotle had long been available in Latin through Boethius.

209   The Grey Friars were Franciscans such as Roger Bacon (d 1294) and Robert Grosseteste (d1253); the Black Friars were the Dominicans (see note 197 above) such as Thomas Aquinas and Albertus Magnus. The recovery of the writings of Aristotle, (of which Albertus knew the entire corpus) gave Scholasticism its character which reconciled the contemporary theology with Aristotelianism. The early university foundations such as Oxford, Cambridge, Padua, and Paris were closely connected with these religious foundations; from whence the design of the Oxbridge Colleges, with their cloisters, chapels, cells and

refectories.

210   Grosseteste was interested in lenses and optics, and he studied the works of Alhazen. As Bishop of Lincoln he strongly pressed for the study of Greek and Latin so as to access the classical sources directly.

Roger Bacon and John of Peckham studied the works of Witelo on mathematics and the optics of Alhazen. Bacon was interested in languages and the collection of scientific data, stressing the virtues of experimentation. His interest in flying, explosives, navigation of the globe and mechanical propulsion were centuries ahead of their time, as was his understanding of convex lenses and their application for spectacles.

John of Peckham was Archbishop of Canterbury. Roland of Parma wrote on medical matters and anatomy, following Roger of Salerno who had benefitted from the Arabic texts in this field.

Leonardo of Pisa, called Fibonacci (1170-1245) learned the Indian numeral system in North Africa and produced his *Liber Abaci,* which underpins how we use number today. The Fibonacci series is more detailed in the footnotes to Canto the Ninth and found to dominate structures throughout the living kingdom, and quite possibly to determine how consciousness creates.(See Penrose's hypothesis in 375 and 376)

211   Nicolas of Cusa (1401-64) preceded Galileo in his advocacy of acute experiments: weighing, changing, observing and weighing again, understanding that plants absorbed elements from the earth and weighing would indicate how much. He supposed that the earth moved (before Bruno) and the universe to be infinite, heretical at the time.

212   The persecution of Christianity led to the formation of remote monastic foundations in the Hebrides and the islands off the coast of Ireland, where rugged weather and rocks protected them. Among the Celtic Christians, the copying of early biblical texts was of paramount importance and their ornamentation became of the highest complexity in design and materials, gold and precious stones. Later Charlemagne (AD 742-814) employed Celtic monks to continue their work under his protection in France.

St Columba, the Abbot of Iona was said to preach to the seals.

Involution

213   Fulda, St Gall and Reichenau were all Carolingian Monasteries under the protection of Charlemagne. Erigena (John Scotus AD810-c877) translated from Greek into Latin and wrote a treatise, *On the Division of Nature*, which was influenced by Neo-Platonism with some aspects suggestive of a kind of pantheism, saved from heresy only by distinguishing the 'Uncreated Creator' from everything else.

214   Cluny was the first and most powerful Benedictine Monastery in the Saône region near Mâcon in France, from which the 'Cluniac' abbeys all derived. Given full autonomy by William 1st, Count of Auvergne, the Cluniac Abbots became very powerful as advisors to the French Courts from the 10th to the 12th centuries. Cluny was sacked many times, disastrously during the French Revolution. The Benedictines were dedicated to the furtherance of learning and established many daughter priories, which were part abbey, part manor, and usually agriculturally self sufficient. They attracted powerful pilgrims and generous donations.

215   Chartres Cathedral, constructed between 1193 and 1250 was, unusually, built very quickly for a high Gothic cathedral. It is known for its outstanding use of flying buttresses which enabled the great height to admit clerestory windows of great size. Light is almost what carves the structure. Originally a Carolingian church of pilgrimage (c 850) it became associated with the Virgin through its acquisition of the '*Sancta Camisa*', the garment believed worn by the Virgin during the birth of Christ. The west portal which opens onto the market square has three entrances, each flanked by a trinity of figures, which Kenneth Clark believed showed, in the 12th century, one of the first recoveries of classical style in the fluted draperies. The cult of the Virgin here also showed a relaxation of the rigid Gothic stance, both more feminine and more universal (and more Greek).

216   Giotto (1276-1336) was a Florentine who was commissioned to fresco the Church of San Francesco in Assisi, where he came directly into contact with the powerful influence of St. Francis, the saint most in touch with the world. His was the first escape from the strict formality of Byzantine representations, (in which line, colour and texture alone depicted almost abstract forms) towards a visual and spatial context and naturalistic portrayal of figures in space with

real attitudes suggesting emotion. The most progressive work, which set Western art on the path towards total naturalism, is in the Arena Chapel at Padua. Giotto was ahead of his time by almost 150 years.

Duccio (1255-1319) was a Sienese painter, a contemporary of Giotto, but still held in the conventions of the Sienese School of Italo-Byzantine formality; Sienna being in closer contact with the East. A painting of the *Three Marys at the Tomb* in the cathedral shows an angel floating off on Christ's tombstone, with little anchorage for the figures or perspective to the tomb. It tries to suggest space but remains a flat decorative panel.

217   Tartini's fiendishly difficult 'Devil's Trill Sonata' came, he said, in a dream when the devil sat on his bedstead and played to him. He claimed he merely wrote it down.

218   Watteau (c 1777) is rather irreverently (and out of chronology) invoked here. He exemplified the extravagant luxury and frivolity of the period of Louis XIV, and painted a number of 'fetes'; 'Girls on a Swing' or extravaganzas *'en plein air'*.

Footnotes to Canto the Sixth

219   Giordano Bruno (1548-1600) was and remains a controversial figure. Born in Nola near Naples, he showed early promise across a range of fields: philosophy, mathematics and astronomy. His theories extended to mnemonics and memory, on which he wrote a guide, *The Art of Memory*. (The pastiche in the text is a travesty of that, but an economical way to indicate the reasons he was a danger to the Church's orthodoxy of the time.) He took early orders with the Dominicans and remained in the Order for eleven years, but perilously.

Caustic and outspoken, he made enemies easily, but his gifts attracted influential patrons at most of the European university cities, Padua, Paris, Geneva and Oxford where he was befriended by Sir Philip Sidney. Usually on the run from indiscretions, he nevertheless wrote extensively on many subjects. His was not an experimental approach but his thought was widely influenced by Arab astrology, Neo-Platonism and hermetic teachings of the Renaissance, notably the Arian heresy which held Christ as 'created' rather than God incarnate. It was the heresy of his theology rather than his scientific ideas that most now feel was responsible for his end at fifty two, but they were inextricably linked with his hypothesis on non-creation, and his refusal to adhere to Aristotelian ideas. He refused to recant.

His views on many things now seem prescient of the 20th century ideas, many now accepted. He believed in an infinite universe, uncreated, in which God was present (immanent) everywhere in his multiplicity of existences (perhaps a forerunner of Everett's 'Many Worlds' theory to explain quantum mechanics). The whole universe was intelligent and moved under its own animism without need for a 'prime mover'. He accepted the Copernican heliocentric universe (as did Galileo) but it was contrary to the Aquinas doctrine of the theologians. More strikingly, the universe was homogenous, composed of the four elements (air, fire, earth and water) all composed of atoms separated by ether (for no vacuum could exist…the 'ether' remained a belief up until Einstein). Our sun was merely one of many stars and the 'units' of the homogenous universe were suns, stars and planets. His was a very modern mind.

His condemnation by the Roman Inquisition (after a seven year trial) led to him being burned at the stake in the Campo dei Fiori in which a statue was erected only in 1889. He was said to have answered his Inquisitors' judgement thus: 'Perhaps you pronounce this sentence against me with greater fear than I receive it'. Prescient indeed!

220   Arthur Wellesley, a career British Army Officer, was the First Duke of Wellington, the hereditary title bestowed on him by George the Third. Other titles were given by the Netherlands, Spain and Portugal for his successes in the Peninsular Wars. He defeated Napoleon at the Battle of Waterloo (now in Belgium) in 1815 with the help of the Prussians, which ended Napoleon's rule and restored the French Monarchy.

221   A quote from Hamlet, 'When sorrows come, they come not single spies, But in battalions'. (Act 4 Sc 1) This is meant to suggest that the weight of discovery and numbers of creative men during the Renaissance is perhaps a reflection of involution's 'recovery' of the explosive natural world during the Cambrian explosion of prolific divergence.

222   This portrait of a medieval tomb envisages the lord and lady of a modest manor parish church carved, recumbent and at perpetual prayer, with their offspring on the sides of the tomb, girls on one side, boys on the other. A kirtle is a lady's long gown or petticoat.

223   Antonio Pollaiuolo (1429-1498) belonged to a group of artists, known as the 'scientific' group, which included Paolo Uccello, della Francesca and Luca Signorelli. Each was rediscovering the secular world in different ways, setting figures in defined space, and accurately portraying features and musculature. Pollaiuolo first made his name as a sculptor, using dissection for his studies, and then using that knowledge in movement, often violent battles between nudes. Formal organisation remains as a decorative element but the explicit detail is new.

Masaccio (1401-1428) was the natural inheritor of Giotto's spatial awareness. He frescoed the Brancacci Chapel in Florence in which scenes are set in natural or architectural landscapes for their context. Vasari claims that Masaccio was the first artist to 'attain to the imitation' of things; he 'made his figures stand upon their feet'. In other works he showed trompe l'œil perspectives, creating 'chapels' giving three dimensional depth to a flat wall framing a painted crucifixion.

Leonardo da Vinci (1452-1519), the ultimate 'Renaissance Man', is seemingly a world away (although only half a century) from the previously mentioned artists. His notes run to five thousand pages of observations on everything under the sun: botany, geology, zoology, optics and engineering, on language, on perspective and on light. In his treatise, *On Painting*, he defines the purpose of painting, 'A good painter has two objects to paint- man and the intention of his soul. The former is easy, the latter hard, for it must be expressed by gestures and movement of the limbs … modelling with light and shadow is the heart of painting'. The quality of his subjects existing somehow 'behind' the canvas was achieved by his use of transparent glazes, so that the light of the canvas illuminates 'through' the subject. Leonardo's designs for a catapult, for a helicopter, for a submarine all appear in embryonic forms, a genius five hundred years before his technological time. Interestingly (for Iain McGilchrist's hypothesis of the creative power of the right brain dominance) Leonardo was left handed and therefore very right brain dominant. His sketches are identified by hatching from top left to bottom right, difficult to imitate, as was his meticulous mirror writing.

224   The Ghibelline and Guelph were factions supporting the Holy Roman Emperor and the Papacy respectively. These allegiances altered, some Italian city states supporting first one and then the other. Most active during the 13th and 14th centuries, skirmishes continued as various city states fell and were resurrected, although the names were used until 1529 when Charles V established his imperial power. They appear in Dante's *Divine Comedy* and in Boccaccio's *Decameron*.

225   The Portuguese explorations of both the New World and the African coast are often attributed to the influence of Henry the Navigator (c 1415) who sought to establish trade routes and protect Portugal from Muslim piracy. In fact the Portuguese exploration had begun a hundred years earlier under King Denis who established

the navy and traded with Flanders under a Genoese Admiral. Under Henry maritime exploration was continuous, expeditions to the Azores, Madeira and the North African Coast were undertaken.

Bartholomew Dias rounded the Cape of Good Hope in 1448 disproving the belief (since Ptolemy) that the Indian Ocean was landlocked. It gave the Portuguese a monopoly on the slave trade for a hundred years. Vasco da Gama rounded the Cape and discovered the sea route to Calcutta in 1498. This was followed by Cabral's discovery of Brazil.

The navigational abilities of the Portuguese were accompanied by developments in navigation, astronomy, cartography, and their development of more adaptable ships, notably the caravel which had two masts and a more flexible sail system allowing it to sail closer to the wind.

Magellan was born Portuguese but sailed under the Spanish Flag. He was the first to sail round the world in his expedition from 1519-1522 and the first to cross the Pacific, named by him 'the Peaceful Sea'.

226 Gneiss rock is a metamorphosed (often banded) rock of either igneous or sedimentary origin. On the Outer Hebrides it forms the bedrock, the oldest in Europe. Gneiss is the oldest rock in the world (3 billion years) and formed during the Pre-Cambrian period.

227 Oolitic is sedimentary rock, formed from ooids, spherical grains, often calcium carbonate crystals, which are formed, usually on the sea floor and which, under pressure, form a continuous sedimentary layer of limestone. Ooids are not unlike kidney stones in structure and formation.

228 James Hutton (1726-1797) was born in Scotland and is considered the father of geology. He made what was then a revolutionary suggestion that the earth had been formed due to the activity of its molten core, and had evolved over millions of years. He added the volcanic cataclysms to the slow formed strata he could observe, which caused stratified rock's upthrust into formations of mountains, which then were eroded. He searched for the evidence and found sedimentary rock penetrated by granite in such a way as to show the granite had been molten when it 'flowed' into the layers. This anomaly became

known as the Hutton Unconformity and conspicuous examples were found in the Cairngorms and at Siccar Point where vertical layers of rock were overlaid by horizontal sedimentary layers, showing the two processes working at different times. Hutton talked of 'deep time' and believed the earth much older than the 6000 years then believed to be its age. His 'Plutonic' views were initially disregarded. The Neptunists, who believed all formations were laid down in the sea bed and the earth's formation was due to a great flood, were, at the time, dominant. He also proposed a theory of evolution by natural selection, as well as climate and rainfall's relationship to humidity. Another genius long before his time.

Charles Lyell (1797-1895) accepted many of Hutton's ideas, indeed made them accessible, but laid less emphasis on cataclysmic volcanic activity. His was the slow cyclical view of earth's formation that largely inspired his friend, Charles Darwin's similar approach to evolution, over similar time periods. It was given the name 'Uniformitarianism', to convey slow changes over aeons of time rather than Catastrophism, the idea of abrupt changes. His 'Principles of Geology' was the seminal work for two hundred years, and much remains current thinking. Stratigraphy was the new specialism that he understood; the measurement of geologic age by fossil contents, and he coined the terms Pliocene, Miocene and Eocene.

229	The Greenwich meridian was developed for a maritime nation, so as to compare the longitude of the position of any ship with reference to the zero longitude, accepted as the 'standard' at Greenwich. One ship's chronometer was always on Greenwich Mean Time, another charted the actual position as the ship travelled. By comparing the two, the position of the ship was calculated (and more accurate map details recorded). The meridian marked the sun's crossing at noon (its highest point) although due to the elliptical shape of the earth's orbit, and the tilt on its axis, it may vary by as much as sixteen minutes, hence the necessity for 'mean'. It was internationally accepted for other nations in 1884, and became the standard 'ahead of GMT' or 'behind GMT' worldwide. Atomic clocks are now more widely used in what is called 'Coordinated Universal Time' which overcomes the anomaly that occurred when the Romans changed Ptolemy's zero at noon to zero at midnight.

230   Alfred Russel Wallace (1823-1913) had observed the differences in distribution of animals and published his book, *Geographical Distribution of Animals,* in 1856, with particular reference to mammals. He divided the world into six regions (obviously selecting mammals would skew the divisions towards later developments but in fact it accords fairly well with perching birds, and certain later developed invertebrate groups).

Lyell's understanding of plate movements and land mass changes had shown, in the fossil ageing of stratigraphic analysis, the underlying explanation of isolation and specialisation of different species (although he did not initially endorse ideas of transmutation from one species to another). Wallace travelled widely throughout the Malayan Archipelago and observed similar but distinct species living in close proximity. In his diary he describes his musings on why there were not the 'intermediaries' and what determined survival of the successful? Malthus's *An Essay on the Principle of Population* came into his mind while in a fever on the island of Ternate, and he saw that the explanation had to be that the 'fittest' survived.

Wallace wrote to Darwin and enclosed a paper on *The Law which has Regulated the Introduction of New Species*; followed by an essay in 1858 which almost exactly laid out the theory on which Darwin had been working for twenty years. On the advice of Lyell and his friend Hooker, Darwin presented Wallace's essay and excerpts from his own at a joint presentation to the Linnaean Society on 1st July 1858. The impression given was that Wallace was a co-discoverer, but he was not party to this decision; too far away to be consulted. Although Darwin behaved honourably it was somewhat at gunpoint in terms of Wallace's intention to publish imminently.

There are differences in emphasis between the two naturalists; competition between individuals drove the pressure for Darwin, whereas for Wallace it was pressure from the environment which shaped adaptation. He saw the animal (or plant) in the context of all about it, and a kind of 'systems theory' explanation of feedback would quickly correct any imbalance or deformation that appeared at an early stage.

The 'Wallace effect' is now widely accepted as an explanation for mechanisms that prevent hybridization between divergent species. Once two populations have become adapted to different environments any hybrid mating would produce offspring less adapted than either parent group. This is considered an important mechanism for stabilising species that originally diverged from a common ancestor. His was a more 'integrated' understanding of mechanisms of natural selection than Darwin's more 'competitive' one.

In 1864 Wallace became a spiritualist and published a paper applying the theory of evolution to mankind. He believed that man had dominantly and rapidly developed due first to his bi-pedalism, which freed his hands, which then permitted the activity of the human brain to render any further physical adaptation almost outmoded. Advantage then became mental which explained the stasis of Homo sapiens as a single species, capable of inter-breeding. Importantly, as a support for *this work,* he believed that evolution could not account for mathematical, artistic or musical genius, humour, wit or metaphysical musing. The 'unseen universe of spirit' had intervened at critical times: the change from inorganic matter to life; from life to consciousness in higher mammals, and from that to intellect had been the result of the intercession of spirit. The raison d'être of creation was the development of spirit. Wallace was open and courageous in the face of disapproval by his peers. In short he more closely fits the 'genius' than Darwin, being broader in his interests: from the politics and radical social issues of John Stuart Mill, Robert Owen and Thomas Paine, to psychological enquiry, and mesmerism.

231   Voltaire (1694-1778) was the great French Enlightenment writer, historian and philosopher, who not only wrote plays, poetry, novels, essays and more than 20,000 letters but touched on almost every subject in the course of his life. Educated by the Jesuits in Latin and Greek he also mastered Italian, Spanish and English in later years. His satirical and trenchant wit caused him to be persecuted in France for his criticisms of the Catholic Church, so in exile in England he studied the English constitutional monarchy, which he much admired, as well as Shakespeare's plays (little known abroad at the time).

He took a mistress, the Marquise de Châtelet, an erudite woman whose translation of Newton's Principia Mathematica was the definitive French translation well into the 20th century, and with her made Newton's ideas more generally understood. Together they studied experimental science, philosophy and metaphysics. His play *Candide* was a satire on Leibniz's optimistic determinism. His great compendium of articles, the *Dictionnaire Philosophique* analysed Christian theory and dogmas, and proposed a new kind of historiography in which social customs, arts and science were more valid indications of historical events than wars and kings. His legacy influenced both the French and American revolutions in its call for freedom of religion, free trade and a universe based on reason and respect for nature. His respect for a universal history crossed all national boundaries, and included Islam and the Far East.

232   Richard Dawkins's many works on evolution; *The Selfish Gene, The Extended Phenotype, Out of Eden, The Blind Watchmaker, Unweaving the Rainbow*, ending with his emphatic attack on religious belief, *The God Delusion*.

233   Biltong is raw cured and dried meat, usually beef or ostrich. It sustained the Boers in the Boer war who carried their protein in a pocket. Light to carry, it does not deteriorate in the heat. It is known as jerky in the United States. 'Bleu' refers to the preference for very rare beef. It may be of interest to some that this author is a member of a family half Boer, half British who had members fighting each other on both sides of the argument!

234   The Trireme is a Hellenistic warship (literally three rows of oars) used by the Greeks and Romans.

235   René Descartes (1596-1650) the 'first modern philosopher' wrote his *'Discourse on Method'* in which he set forth the standards to be attained by the building up of knowledge, eliminating tradition and belief and starting with thinking itself. It was said that while lying in bed and observing a fly hovering in the corner it occurred to him that at any moment its position could be defined by its distance from the three planes formed by the adjacent walls. These three could be reduced to two 'Cartesian co-ordinates', thereby seeing a curve as a moving point. An equation between these two could then describe the property of the curve. In this way he re-introduced the correspondence between number and form as understood by Pythagoras and Plato (later separated in Alexandria and studied in isolation). He introduced movement into geometry, and this was also the field explored by Galileo.

His scientific legacy was more profound than his discoveries, for it was from Descartes' *Meditations* (1641) that thought was severed from the created world, which persists to this day. 'This I- that is the soul by which I am what I am- is entirely distinct from the body... and is easier to know'... 'I think therefore I am' hardly does justice to 'I am only what my thought makes me'. It contributed significantly to science's subsequent indifference to the organisms that were reduced to what could be known of them by man, mostly by killing them. The analytical method sanctioned by Descartes has run through science ever since.

Yet thought experiments have contributed significantly to intellect's analysis and conjectures; from Kepler's view of the Earth from the moon, Einstein's journey on a beam of light and perhaps to the Shamanic belief in the validity of thoughts communicated by plants and animals.

236   Blaise Pascal (1623-1662), a friend of Descartes, developed probability theory and new geometrical concepts. He intuited that resistance was due to a vacuum in a column of water, and investigated the effects of altitude on barometric pressure. The idea that air has 'weight' would later be developed by chemists realising that air consisted of oxygen and hydrogen in fixed proportions which led to the advances in atomic theory.

237   William Gilbert (1546-1603) was the author of *De Magnete - On the Magnet and on Magnetic Bodies and concerning that great Magnet, the Earth, a New Physiology*, which was admired by Galileo. He realised that at the poles a free needle would point vertically, at the equator horizontally, and at intermediary positions between the two. This he explained as due to the magnetic field of the earth itself. He was, with Bruno, also one of the first to propose that the so called 'fixed stars' were all at differing distances from the earth.

238   Galileo Galilei (1564-1642). It is impossible to do justice to the breadth and versatility of Galileo's enquiries, studies or observations. What made him the father of modern science were not merely his discoveries but his ability to perceive in the smallest phenomenon the relationship of underlying law, applicable to the whole universe. Initially a student of Aristotle, within five years he had seriously undermined much of the Aristotelian universe. His demonstrations

on falling bodies (1lb and 100lb fell at the same speed) showed that the rate of fall was not a function of weight. This implied a force acting equally on both. He distinguished between primary and secondary qualities, the latter susceptible to the senses and therefore subjective and unreliable. From this follows 'Science as Measurement' which persists to this day.

His '*Dialogue*' presents the uniformity of the material universe; the doctrine of uniformity (causes produce corresponding effects) is fundamental to all scientific disciplines. From Galileo we get the mechanical world which affected every science, from physics to chemistry and biology, with his mechanical principles applied to the earth, the heavens, the minute and the very large, as well as to the bodies of animals.

Il Saggiatore (The Assayer) is an allegory Galileo wrote on the motives, methods and limitations of the 'experimental philosophy' he knew he had created. It gently satirises the over confident experimental destruction of the very explanation being sought. As an example he recounts the inability of an observer to locate the song of the cricket, but in his attempt killing it, and thereby still failing to get an answer. This allegory is extended in this passage to mean the destruction of any explanations when too firmly gripped, too exactly defined. (It seems an anticipation of Gödel's Incompleteness Theorems, which demonstrate that no system of axioms based upon number theory can prove its own consistency within the system itself. Something always remains outside and unproveable.)

Galileo's tools, microscope, telescope, and thermometer brought in the skilled mechanic as essential to science….extending the range and acuity of the senses, and forging science's new marriage to technology. Our last tool, the computer, is the measure of the recovery of the contents of mind, and like the earliest which accompanied the separate senses, it similarly extends the capacities of mind as the essential tool for analysis of information. Tools accompany and testify to the re-penetration of memory. Now the use of computers to analyse overwhelming data has produced the evidence for Chaos theories' patterns underlying seemingly haphazard cataclysms.

From Galileo we get the boundless mathematical universe, which implies no beginning in time and no limit in space. With him also continues Platonic/Aristotelian split between natural philosophy with its impervious laws and moral philosophy which survives to this day, in the divorce between science and religion, including science and ethics.

Galileo launched the Scientific Age and his precepts continue to govern its approaches, and its repudiation of anything (like this work) seeking to defy or stretch those precepts!

239   In 1638 Galileo, not long before his death and almost totally blind, was visited by John Milton who, sixteen years later, recorded the visit (and what he had learned upon it) in *Paradise Lost*. '*….the moon, whose orb / Through optic glass the Tuscan artist views / At evening, from the top of Fesole, / Or in Valdarno, to descry new lands / Rivers or mountains, in her spotty globe.* (Paradise Lost 1: 286-291) It is interesting that Milton refers to Galileo as an 'artist'. Unlike modern scientific writing Galileo's language is poetical, and Arthur Koestler refers to it as 'literary masterpieces which had a lasting influence on the development of Italian didactic prose'. Koestler goes further 'In our present century Eddington, Jeans, Freud, Kretchmer, Whitehead, Russell, Schrödinger….gave convincing proof that works on science can also, at the same time, be works of literary art…' whereas…the overloading with (scientific) jargon, the tortuous and cramped style, are largely a matter of conforming to fashion'. (p265 in *The Act of Creation'*) I rest my case for this attempted alternative.

240   Johannes Kepler (1571-1630), an exact contemporary of Galileo, could not have been a more different character. Where Galileo had an acute incisive observing mind, (Aristotelian in bent) Kepler was quintessential Platonist, searching for a unitary harmonious structure in the universe, which would be expressible mathematically. Between the two of them the universe was indeed unified as obedient to the same laws of motion throughout.

Born in relative poverty in Weil der Stadt near Stuttgart, his father was a mercenary who disappeared when he was five; his mother, a healer and herbalist, kept an inn and was later tried as a witch, but released. Kepler's mathematical gifts were apparent early, as was his interest in astronomy. Sponsored to attend a grammar school and a Latin school, Kepler was admitted to the Tubingen Stift where he studied philosophy and theology. He was known as 'the frail' due to the damage left by smallpox on his eyesight and his partially crippled hands.

Like Galileo, Kepler accepted the Copernican heliocentric universe and held the sun to be the source of all energy. In his *Mysterium Cosmographicum* he set out a structure based upon the five platonic solids (shapes of equal sides and equal angles) which he believed held the orbiting plants in shells at specific distances from the sun. It turned out to be wrong but it represented the first attempt to give mathematical structure to the universe as a whole.

He suggested that 'there is a moving intelligence in the sun that forces all (planets) round, but most the nearest, languishing and weakening in the more distant by attenuation of its virtue by remoteness'. The words are those of the Middle Ages (intelligence, virtue, languishing) but the concept anticipates gravitation and Newton.

He joined Tycho Brahe in Prague and became his literary legatee, publishing the *Rudolphine Tables* which described the position of 1005 stars. His observations in Brahe's observatory on the planet Mars led to the publication of his *New Astronomy with Commentaries on the Motion of Mars* in which he set out his first two laws of planetary motion. 1. Planets move round the sun in ellipses with the sun at one focus of the ellipse. 2. A line drawn from the planet to the sun sweeps out equal areas of the ellipse in equal times. His third law followed later. 3. The squares of the period of revolution are proportional to the cubes of the distances from the sun.

Two thousand years after Apollonius had described cones, ellipses, conic sections, and parabolas which to the Greek mind were independent of any natural occurrence of them (except humanly constructed) Kepler found them all in planetary motion. This begs the question which this poem attempts to answer:

'Is mind the author of, or attuned to the natural world?'

Kepler's Laws became more widely applied by Newton, which rephrases the same question in another form. 'Are the midwife minds of genius linked together to deliver the natural world?'

241   'Its all rubbish' is its essential meaning.

242   Arthur Koestler's *Act of Creation* is an exhaustive study of both the circumstances in which creativity arises, and defines what qualifies as creativity. It ends with an Appendix (the 3rd) summarising the qualities common to scientific genius. The first is precociousness: for every Mozart in the arts there are three Pascal, Maxwell, and

Edisons in science. Regiomontanus(1436-1476) published the best astronomical yearbook for 1448 at the age of twelve; Pascal had laid the treatment of conic sections before he was 16; Clerk Maxwell read a paper proclaimed superior to Descartes or Newton, to the Royal Society at fifteen.

A second common thread is the absence of conventional schooling, or an indifference to the substance of an education in favour of another passion. Einstein was never interested in mathematics and the conventional physics that he was taught caused him to 'switch off' science for almost a year.

'A student's matrix of thought is still fluid.' Einstein likened the determined appetite of youth to the 'the voraciousness of a beast of prey'. The genius has never lost the habit of asking foolish questions.

Abstraction and Practicality: Scepticism towards the conventional creates a 'freshness of vision, and acuity of perception… the co-existence of concrete and abstract thought; high flights of theory are balanced by a keen sense of the practical…a gift for seeing the potential in trivia…a head for generalisations and an eye for minute particulars.'

Finally: Multiple Potentials: Dr Johnson defined genius thus: 'True genius is a mind of large general powers, accidentally determined to some particular direction, ready for all things, but driven by circumstances for one.' In this view creativity is convertible energy, dealing with problems as wide apart as colour-theory and celestial mechanics (Newton) or electro-magnetism and theory of gases (Maxwell). In the world of the scientific genius, one-idea men like Copernicus and Darwin are the exceptions. So much so, as to raise the question as to whether they qualify. It is interesting that both shared an unwillingness to publish and be judged. Was that because they themselves doubted their contribution? That hesitancy is unusual in the world of genius.

Koestler ends his book by referring to 'the ominous trend to over-specialisation which has dangers for the creative mind'.

243 Multiple Discoveries. In 1922 Ogburn and Thomas published one hundred and fifty examples of discoveries made independently by several people. Later Merton stated that 'the pattern of independent multiple discoveries in science….is the dominant pattern rather than a subsidiary one'. Lord Kelvin published thirty two of his discoveries

which he subsequently found to have been made by others.

Koestler states that the 'ripeness' for new advances seem to be 'in the air'. Einstein in his relativity did not rely on anything that had not been known for fifty years. Well known examples are Cavendish and Helmholtz; Newton and Leibniz on the differential calculus (used by Galileo but not recognised as such); Russel Wallace and Darwin in evolution (both having read Malthus); the independent discovery of Neptune (on the basis of theoretical prediction) by Couch Adams and Le Verrier, and simultaneous insights by Galileo and Hooke. *"When the time is ripe for certain things they appear at different places in the manner of violets coming to light in early spring"*. This appeared in a letter from Farkas Bolyai to his son Janos to urge him to publish his discovery of non-Euclidean geometry. Involution as a hypothesis explains that inevitability. Among them is the simultaneous hypothesis of the Higgs boson in 1964. In terms of evidence for the convergence of involution the lists of such multiple 'discoveries' hugely increase; from the few in the 17th century to the almost universality of them in the 20th century. Genius appears to be becoming increasingly infectious! Is it because we are collectively getting closer to the 'quantum vacuum' binding all?

244   String players put the 'bowings' into a score before playing it, to achieve the synchrony of 'up' and 'down' bows throughout the section during performance.

245   Pope Julius 2nd commissioned Michelangelo to paint the Sistine Chapel, and Raphael to decorate the papal apartments. The brave demolition of the old St Peters and the erection of the new was his decision, as was his faith in Bernini who designed much of the flamboyant sculpture and the Great Court with its double colonnades. Bernini's *St Theresa in Ecstasy* seems overtly sexual in its rapture to those unfamiliar with accounts of spiritual surrender.

246   Raphael's *School of Athens* (1509-1511) is a fresco decorating the papal apartments in the Vatican, and it was painted at the same time as the Sistine Chapel ceiling.

247   Sandro Botticelli (1444-1510) was largely employed by the Medici to work on classical mythical and secular subjects. 'My fancy

is never checked: as the zephyr it flows smoothly along the gull-like pattern of waves on the green sea.....You soon forget the picture... it is so evanescent and shy, as rare as a dream'. In tandem with the individual scientists, artists of this period explored different aspects of realism, each making separate contributions to the growing portrait of collective mind, now absorbed with the world of the senses, and the freedom of reason.

248   Albrecht Dürer (1472-1528) largely escaped the strict Gothic traditions of German art, and was quickly alive to the reproductive possibilities of the new printing press. He developed both woodcut and etching as a response to the demands for illustration. His etching shows extraordinary detail of surface textures, wood, metal, fur, and glass simply in gradations of grey. Earlier than other Renaissance painters in Italy, his was a voraciously enquiring mind, although very self absorbed. His self portraits painted throughout his life (the first at 13) show a remarkable face, with startling, rather cold, penetrating eyes.

249   Giorgione (1478-1510) left few paintings (as he died at about thirty) but used colour very skilfully to convey aerial perspective: warm browns and reds against cool blue and green. In his *Fête Champêtre*, under a lowering stormy sky, an informal group demonstrate his creativity with movement, not only the figures of the nude women, but the animated talk of the clothed men, and the secular composition as a whole is freed from the formality of architectural references or prudishness.

250   Titian (1477-1576) initially collaborated with Giorgione and used similar techniques of rich colour masses built up in many layers. Apart from mythical subjects, he painted a number of startlingly revealing portraits of great simplicity in structure and subtlety of character, highly individual. Later he painted with emphatic chiaroscuro, to capture the drama of heroic subjects, often biblical in origin, such as his *Entombment*, painted with such dramatic economy they become universal. Even commissioned religious subjects were now giving way to extreme naturalism.

251   Jan van Eyck (1385-1440) was an earlier Flemish painter who

resided for most of his life in Bruges, widely recognised even in his own day. His early work was influenced by the tradition of miniature paintings prevalent in Flanders (where, unlike Italy, there was an absence of great wall spaces) and this tradition finds its way into the jewel-like paintings, even of secular subjects like the famous *Giovanni Arnolfini and his Wife*, in which the minutiae are faithfully recorded, the chandelier, the convex mirror reflecting the distorted window, the fur on the gown; the realism of his observations subordinated to the calm static composition. Often called the 'father of oil painting', he did not originate the use of oils but developed it to a new level with wet-on-wet glazes to achieve its jewel-like surface. Unusually, van Eyck was well remunerated throughout his life at the Court of Philip the Good of Burgundy.

252   'Nullius in verba' is the motto of the Royal Society, which was the first Academy of Science established anywhere. Members in its early days included Christopher Wren, Robert Hooke, John Evelyn, and Francis Bacon. It literally translates as 'Take nobody's word' and from the outset showed a determination that its findings and research demanded experimental evidence. That, probably more than anything, prevents science from heeding psi phenomena despite widespread evidence of them. Since a central tenet of *this theory* is the inability of the intellect to penetrate consciousness, let alone measure what is infinite, suggestions that psi phenomena should be subject to such experimental procedures would be self defeating; they are precluded by their nature from science as traditionally practised.

It is curious that Rupert Sheldrake, a strong critic of science in this regard, continues to suggest statistical and experimental methods to make psi phenomena more accessible, and thereby recognised. As most geniuses attest, it is only in the abeyance of the will that intellect penetrates consciousness. Such methods may gather limited evidence from the behaviour of animals but there is not a dog owner who could not match statistics with more illustrative anecdotes. Science will be immune to the wider evidence of the mystics. The ego-will acts as a barrier, particularly the sceptical will of statisticians. One of the hopes of involution is that it should help to banish science's belief that it alone is equipped to judge the experiences of others, given the evidence that such experiences have fuelled its own progress and yet failed to be even acknowledged. (See Appendix and the rare example of William

James in heeding the subjective accounts of religious experience of others.)

253   The phrase 'The Ghost in the Machine' was coined by Gilbert Ryle (1900-1976) in his critique of Cartesian dualism. It was used by Arthur Koestler as a title for his book in which he accounts for the levels influencing mind, seemingly autonomously, being due to the Holon: the mind is at once a whole and a part of several levels of inheritance, (themselves equally holons) as well as susceptible to the 'life signals' of the entire group to which it belongs. The 'illusion of autonomy' in the individual is the 'ghost'. In this metaphor that 'ghost' is extended to the whole of creation in which there is seemingly no longer any need for God, the Celestial Mechanic, or a Prime Mover.

Footnotes to Canto the Seventh

254   'Materia' is derived from 'mater' – mother. In medieval and alchemical nomenclature the materia prima was the primal substance of which all things were made; the primal unity that lay behind the four elements.

255   Isambard Kingdom Brunel (1806-1859). A ground breaking engineer, he revolutionised the construction of bridges, tunnels, buildings, railways and ships. Son of a Frenchman, educated in the classics in France, his designs were not only innovative but elegant. His Clifton Suspension Bridge had the longest span of any bridge of the time and is still in use, as are the bridges constructed for the Great Western Railway. He hoped to extend the latter to New York by building steamships running on coal which at the time was considered unfeasible. His great ships were revolutionary: The Great Western was built mainly of steel-reinforced wood, followed by the steel hulled ships, the Great Britain and the Great Eastern which carried passengers in great luxury. He also designed pre-fabricated hospitals for the Crimea and his 'Railway Village' to provide for the needs of railway workers .This is believed to have influenced Aneurin Bevin's creation of the National Health Service.

Brunel exemplifies, in practical terms, the versatility and eclecticism of liberated thinking. He could turn his mind to any problem of construction.

256   An 'eccentric' in geological terms is a stone far from its created origins, often carried by melting glaciers. This description of Newton's appearance is taken from two portraits of him painted during his lifetime by Godfrey Kneller.

257   Isaac Newton (1643-1727) was born after his father's death to Hannah Ayscough in Woolsthorpe Manor, Lincolnshire. He was so

small she said he could be fitted into a 'quart-jug'. She remarried, a hated step-father, and he was left in the care of his grandmother but on her return (after her second husband's death) she tried to persuade Newton to farm the Manor.

'Hypotheses non fingo –I invent no hypotheses'. This appears at the end of Newton's *Principia Mathematica*. It continues '... whatever cannot be deduced from the phenomena should be called an *hypothesis*'. He uses the word here in its Platonic sense, meaning a postulated plan which must be accepted before discussion takes place, as in 'The Pope is infallible' or 'Man is innocent until proved guilty'. In terms of the scientific use— a generalisation drawn from observed data which will further be confirmed—he constantly invented and employed hypotheses.

258   Kepler's laws of planetary motion were both proved and mathematically described by Newton in more general terms. Galileo had come very close to the Newtonian addition of the concept of universal gravitation in which Newton was aided by Edmond Halley.

259   Calculus extended Descartes in its ability to mathematically describe and predict movement in curvature, which had become necessary in answer to the much more accurate observations of astronomy. After Leibniz had visited Christiaan Huygens he devoted years to the development of the differential calculus and the claims of both Leibniz and Newton to primacy raged for the whole of Newton's life, in an embittered quarrel. The species apple tree under which Newton was allegedly assaulted by gravitation was called 'The Flower of Kent'.

260   The shape of the earth and the moon had been demonstrated by LaGrange (1736-1813) as not perfectly spherical, which fact meant that Newton's assumption, that the force of gravitation exercised between them acted at the centre, had to be adjusted. LaGrange distinguished between *periodic* disturbances and *secular,* the latter depending upon orbital size, shape, and planes of movements. These became the *elements* of a planet. Laplace extended these ideas in his *Celestial Mechanics* to describe the stability of the solar system with a fixed quotient of eccentricity and further postulated that the whole universe had condensed from a rotating atmospheric nebula. That

idea persists today.

261   The aberrations, chromatic and spherical, of light passing through a telescope led Newton to examine the different refractions of different colours (separated by passing white light through a prism). He showed that colours were differently refracted, with blue more refracted than red and yellow and green least of all. The eye is most sensitive to those least refracted. Newton showed that white light was composed of this spectrum (and thereby overturned Aristotle's view that colours were white light admixed with darkness to varying degrees). Descartes had assumed that colour was the product of the different rotation of 'corpuscles', and Newton accepted this hypothesis.

  Newton's concept of colour as well as his demonstrations on aberration led to major advances in telescopes. Dolland, following Euler's suggestions, devised the achromatic lens, (correcting the differentials in refraction) that resulted in observatories built first in France where Huygens worked on his famous Horologium Oscillatorium. From an analysis of gravitation's influence on the pendulum clock, came the wider understanding of the principles of dynamics. It seemed to escape Newton, (who believed an achromatic lens to be impossible), that he was constantly looking through two of them!

262   When Newton was offered a fellowship at Trinity College Cambridge, it was assumed he would take holy orders as a priest, a mandatory requirement. He refused. Later when he was appointed to the prestigious Lucasian Professorship (now held by Stephen Hawking) he again refused and was exempted by a privilege conferred by Charles the Second.

263   In 1696 Newton was made the Warden of the Royal Mint, a sinecure which he took very seriously, assuming responsibilities for the coinage, and introducing the gold standard rather than the silver. At the time it was estimated that 20% of coins were counterfeit. To combat this erosion Newton had himself made a Justice of the Peace in all the Home Counties, and having done much of the gathering of evidence himself, successfully prosecuted many counterfeiters (coiners).

264   The Rosicrucian movement has had varied and fitful incarnations. Thought to have been founded by Christian Rosenkreutz in late medieval Germany as an order of mystic doctors pursuing the hermetic path to enlightenment, it had the support of early scientists (Kepler, John Dee, and Tycho Brahe). It later formed the Invisible College, the precursor to the Royal Society, where members met in secret to exchange experiments in alchemy and the integration of science and the arts. The transformation of a hermetic society into the bastion of science is interesting. In his later life Newton pursued Rosicrucian ideas and mixed socially with like-minded others. Later it had Albigensian links. Essentially Rosicrucianism posed problems for the Orthodox Church in its emphasis on esoteric knowledge accessible to the individual.

265   François Xavier Tourte (1747-1835). His father was a bow maker and luthier but the Tourte violin bow devised by François was adopted through its use by the great virtuosi, Viotti and Paganini. France has since remained the home of great bow makers, whereas Cremona in Italy was the centre for the violin making by the families of Stradivari, Amati and Guarneri.

'Sostenuto', a musical instruction means 'sustained; 'spiccato' means short and bouncing, and 'espressivo' means emotionally 'expansive'.

266   Pau-Brazil was the original Portuguese name given to the tree, Caesalpina echinata from 'pau' (stick) and 'braza' (ember-like), a reference to its deep red colour. Originally valued for its red dye, it was a precious source for Renaissance clothing and for the Portuguese its discovery in 1500 gave a great and protected source of income. It only flourishes in mixed forests and is now in such short supply that bow makers world wide, whose livelihood depends upon it, have banded together to supervise replanting. Pernambuco is the name derived from the province in Brazil where it grew abundantly.

267   Spillikins is the game of letting a clutch of fine sticks fall, and having to pick them up one at a time without moving any other.

268   Karl Linnaeus (1707-1778) was a Swede who devised the botanical Latin for the classification of plants and animals that survives to this

day. He took the structural parts in a sequence and grouped them into Species, Genera, Orders and Classes. Plants were classified by the parts of their flowers: one stamen became Monandria, two Diandria and so on.

Animals were classified into the Classes: Mammals, Birds, Reptiles, Insects and Vermes. He was the first to adopt the binomial nomenclature; the first name defined the genus, the second the species. Cuvier adopted a different system on internal analysis of functions (more reminiscent of Aristotle) and a combination of the two systems has gradually been adopted as genetics is more widely applied.

Dmitri Ivanovich Mendeleev (1834-1907). Prior to 1850 it was believed there was no limit to the numbers of elements (elements and molecules were not distinguished: chemistry and physics applied different criteria). In 1864 John Newlands proposed that if elements were arranged by atomic number they demonstrated recurring chemical properties in groups of eight, which he called the Law of Octaves. This was reminiscent of Kepler's music of the spheres and the harmonic scales in western music. Mendeleev repeated this process but left gaps in the order, making the properties the dominant criterion. He predicted the gaps would be filled and when he discovered the elements Gallium, Scandium and Germanium his insight was rewarded. Rather in parallel with biological classification, elements indeed fell into families such as alkalis, halogens etc. The periodic table has been filled with rarer elements since Mendeleev.

269   The Sargasso Sea is the only sea without shores. It is known for its rich blue colour and exceptional clarity. Bounded by the Gulf Steam to the south, the North Atlantic current to the north and the Canary to the east, it is a rich depository of a wide variety of organic material from all directions which gets trapped in the gyre. It was named by the Portuguese navigators after its covering of Sargassum seaweed, but there are records going back to navigation in the 4th century BC. It has always been a sea of great mystery and galleons believed to be marooned for centuries have appeared intermittently in literature. '*The Wide Sargasso Sea*' was the title of a novel by Jean Rhys.

270   Robert Boyle (1627-1691). Born in Ireland, educated at Eton and Oxford, Boyle visited Galileo, and worked closely with Robert Hooke on the Guericke air pump examining the properties of air, its

elasticity, compressibility and weight. He showed that the volume of any gas varies inversely to the pressure upon it (Boyle's Law) and in his *Skeptical Chymist* went further in proposing the existence of elements, '...certain Primitive and simple bodies which not being made of other bodies, or of one another are the Ingredients of which all mixt Bodies are compounded and into which they are resolved'. In his *Origin of Forms and Qualities* he 'espoused an atomical philosophy' in which all matter consists of particles.

His experimental approach was wide ranging and covered the propagation of sound through air, the expansion of ice, the nature of specific gravity, crystals and electricity. He distinguished between elements and clusters (compounds) and together with Hooke recognised that only part of air was used in respiration, anticipating the discovery of oxygen. Between Boyle and Hooke the foundations of almost all contemporary practical sciences were laid down.

Like Newton he refused to take orders, but was a keen promoter of Biblical translation so that people could study it for themselves.

271   John Dalton (1766-1844) extended Boyle in realising that gases expand equally with equal increments in temperature. (Gay-Lussac had already described this.) In studying air, Dalton realised it combined oxygen and nitrogen in fixed proportions (with small quantities of water vapour and carbon dioxide). From this understanding– that these gases were not in combination and had different densities– he proposed that their distribution suggested a constant atomic movement of distinct substances in fixed proportions. He started with the assumption that chemical combination took place in the simplest possible way, one atom of one combining with one atom of another. This was in fact wrong (water is two H to one O) but it led to molecular theory and the calculation of relative atomic weights.

It was Avogadro who derived molecular (little masses) theory, arguing that Dalton's atomic 'clusters' were not indivisible but could be separated in chemical reactions.

272   Occam's razor, the principle attributed to the logician William of Ockham (1285-1349) is the law of parsimony (*lex parsimoniae*) which recommends selecting the hypothesis which makes the fewest assumptions about the available data; the razor 'shaves away unnecessary assumptions'. The contemporary multiverse theories with

ten or eleven dimensions, all unproven and invisible are a flagrant defiance of Occam's Law, which has prevailed in scientific reasoning for centuries. Perhaps desperate times (the insufficiency of matter) permit desperate theories.

273   Niels Bohr (1885-1962) was the Nobel Prize winning Danish physicist. He worked under J.J. Thompson at Trinity Cambridge and with Rutherford, whose ideas on nuclear structure he adapted to take account of Max Planck's quantum theory. He devised the model of the atom, in which electrons orbit around the nucleus, and the electron number determines the chemical properties. Light, in the form of photons, was emitted when an electron dropped from a high energy orbit to a lower, which formed the basis for quantum theory. Bohr's Principle of Complementarity asserted that contradictory properties (such as light being a stream of particles, or alternatively a wave) could be separately analysed and equally validly 'true.'

274   Robert Hooke (1635-1703), a close colleague of Boyle, was a polymath of wide ranging abilities, widely admired by Christiaan Huygens and Wren. As Curator of the Royal Society he was at the forefront of all new emerging ideas, studying gravity, barometric pressure, the differences between venous and arterial blood, as well as coming close to declaring the relationship between gravitation and the inverse square law. He had a fractious relationship with Newton.

John Locke (1632- 1709), a philosopher and physician, developed the *Theory of Mind* which was influenced by Descartes. Mind was initially a 'tabula rasa' on which was inscribed the consequences of reason derived from the senses…'the self is defined through the continuity of consciousness and the human is distinguished by the self-reflective consciousness, and toleration'. He was an influence on Thomas Jefferson and Voltaire in his wide ranging enlightened liberalism.

275   This extended metaphor of the loom has some technical terms: The warp is formed by the vertically structured threads through which the horizontal weft is woven by the shuttle. The fell is the body of the growing fabric and the selvage is the strong edge at either side.

Involution

276   Krakatau (pronounced Krakatoa by early explorers) is an island between Java and Sumatra. The largest volcanic eruption recorded in 1883 destroyed a third of the island and 40,000 lives.

277   Alexander von Humboldt (1769-1859) was a widely travelled observer of the patterns across the globe, including the distribution of temperature and pressure, which led to his matrix of isothermal lines, and provided the basis for meteorology. He showed that volcanoes occurred in linear groups which he supposed corresponded with subterranean fissures, and evidence for this was to be found in igneous rocks previously believed aqueous in origin.

Abraham Werner (1750-1817) taught at the Freiberg School, hardly travelled, but was the founder of the gradualism of the Neptunists who believed that all rocks were laid down in steady succession of sea deposits. Known as an intolerant didact; his ideas influenced Darwin's similar incremental gradualism in evolution.

278   Gay-Lussac (1778-1850) extended and applied the ideas of Robert Boyle, and the two were later linked by Avogadro's molecular theory.

279   Ramsden's (1732-1800) equatorial can be adjusted to cause a telescope to follow by clockwork an apparent motion of any point in the heavens. Harrison's (1692-1776) chronometer led him to be called 'Longitude Harrison'. His chronometer enabled an exact determination of longitude by suspending a clock on gimbals enabling its absolute horizontal position to be maintained. It kept time to within 54 seconds over a five month period. Initially made in response to a competition to solve the problem of calculating longitude (the angular distance from the Prime Meridian at Greenwich), so as to plot sea voyages accurately to avoid and map dangerous coastlines, Harrison was very ill-treated by the Board (despite the accuracy of his invention) and required the influence of George the Third before being given the money he was owed. *'The lonely sea and the sky, And all I ask is a tall ship/ And a star to steer her by'* …John Masefield's evocative poem.

280   Little Black Sambo was a children's story by Helen Bannerman first published in 1899. It depicts a South Indian Tamil child who encounters four hungry tigers. He surrenders his new clothes, shoes

and umbrella so they will not eat him. The tigers chase one another round a tree for the possessions, until they are melted in a ring of butter. Sambo recovers his possessions and his mother makes pancakes with the melted butter. This analogy refers to the reductionism of science that slices up living creatures believing that the whole organism can be understood by analysing the parts, killing life to understand life. In the footnotes to Canto nine, recent breakthroughs in understanding are detailed which have come from studies in living cells, or molecules (in vivo) not evident in 'in vitro' laboratory studies. As is often the case, life trumps dissecting analysis.

281   Wolfgang Pauli (1900-1958) was an Austrian theoretical physicist, a pioneer in quantum theory. He articulated what came to be called the Pauli Exclusion Principle which stated that no two electrons could occupy the same quantum state (defined by the four quantum numbers plus the two degrees of freedom). The latter came to be called 'electron spin'. In 1930 he proposed the existence of a neutral particle with very small mass which Fermi later called the neutrino- the little neutral one. (Its existence was confirmed two years before Pauli's death.) Shortly after this he suffered a breakdown and became a client of Carl Jung who derived many of his ideas on archetypes from Pauli's vivid dreams.

In parody of an undetectable neutrino John Updike wrote this verse. 'Neutrinos they are very small/ They have no charge and have no mass/ And do not interact at all/ The earth is just a silly ball/ To them, through which they simply pass/Like dustmaids down a drafty hall/ Or photons through a sheet of glass/ They snub the most exquisite gas/ Ignore the most substantial wall/ Cold shoulder steel and sounding brass/ Insult the stallion in his stall/ And scorning barriers of class/Infiltrate you and me! Like tall/And painless guillotines they fall/ Down through our heads into the grass.'

282   Bartolomé Esteban Murillo (1617-1682) was a Spanish Baroque painter born in Seville, whose early charming genre paintings of urchins and peddlers later gave way to his sentimentalised Madonna's and the religious themes required for his commission to decorate the monastery of St Francisco el Grande.

Francisco Goya (1746-1828) called the 'last of the old masters, the first of the moderns' influenced both Picasso and Manet. His early paintings as Court painter to Charles IV required innumerable portraits (the Royal Family portrait was described by Gautier as the 'corner baker and his wife after winning the lottery'). His later, much darker works were savage exposures of the Peninsular War battlefields, (the Disasters of War) and his series on prisons and the treatment of the insane (the Black Paintings) were savagely grotesque.

El Greco (1541-1614) was a Spanish Renaissance painter born in Crete who after initial work in Venice and Rome settled in Toledo for the rest of his life. His painting received perplexed reactions at the time due to their mannerist proportions and bizarre perspectives but later influenced Impressionism.

Diego Velasquez (1599-1660) spent the entirety of his life as Court Painter to Philip IV with a continuous demand for portraits of court characters; wives, dwarfs, buffoons. Commissioned by Philip to establish an academy in Spain he travelled throughout Italy visiting painters and buying paintings; Titian, Tintoretto and Veronese, as well as painting his famous subject in Rome, Pope Innocent X. His most perplexing painting is called Las Meninas, because its real subject is debatable. Is it the Infanta and her companions? Is it the view from the mirrored figures watching, or is it the painter half hidden behind the canvas and also portrayed? All perspectives are simultaneous. His beautiful Rokeby Venus was the first full length secular nude ever painted.

283   Jackson Pollock (1914-1956) was an American painter famous for his drip paintings, using liquid paint. He painted these large canvasses on the floor from all directions so as not to impose the single dimension of the vertical canvas. In later years he refused to title his works for the same reason, they were given numbers to avoid any overt suggestion as to meaning or subject. He wrestled with alcoholism and his fiery nature all his life, and was killed in a car accident under the influence.

Piet Mondrian (1872-1944) was a Dutch painter who was an influence in the establishment of the De Stijl movement in art which was later called Neo-plasticism. His early work followed The Hague School of impressionism, largely landscapes, but increasingly moved towards a restricted primary colour palate. He was, throughout his life, in pursuit of spiritual integration; a follower of the Theosophical Society started by Madame Helena Blavatsky. Briefly, during work in Paris before the war, he was influenced by the cubism of Braque but the works for which he is best known are the canvases of black lines on white ground, with only red, yellow and blue shapes, increasingly spare.

He wrote ' I believe it is possible through horizontal and vertical lines, constructed with awareness, but not calculation, led by high intuition, and brought to harmony and rhythm, these basic forms of beauty…can become a work of art, as strong as it is true'.

For Mondrian painting was a kinetic meditation on the 'essence' stripped to a minimum.

284   Peter Paul Rubens (1577-1640). A Flemish Baroque painter renowned for his colour, movement and sensuality. Educated in the classics, his early work was commissioned by Vicenzo Gonzaga in Mantua, for whom he also acted as emissary, travelling to Italy (where he encountered the work of Titian and Caravaggio) and to Spain. In Italy he painted several portraits which were to influence Van Dyke, Reynolds and Gainsborough. On the death of his mother he returned to Antwerp where he concentrated on large altarpieces. After the death of his first wife he married the voluptuous sixteen year old Hélène Fourmant who appears in several mythological scenes as Venus etc.

285   Antony Van Dyke (1544-1641) was an almost exact contemporary and pupil of Rubens and the leading Court painter to the English Court (after his training in Antwerp and Italy) who painted over forty portraits of Charles 1st as well as members of his family and other English nobility. He was a pioneer in water colour and etching and his full length portraits have a new relaxation and elegance with an undiminished authority. This 'English Style' of informal portrait would be continued by Reynolds and Gainsborough. Many of his large works were finished by assistants brought from Antwerp because there was no school of training in England. Hence '*Painting could be mastered somewhere else*'.

286   Rembrandt van Rijn (1606-1669), the apogee of the Dutch Golden Age, extended the role of the painter to every genre and in a sense 'secularised' biblical subjects so that the biblical message becomes renewed and relevant. His early portraits made his name initially, but recognised and fostered by the statesman, Constantijn Huygens (the father of Christiaan), he received commissions from

prominent people in The Hague. His innumerable self- portraits give a biography of the man from initial wealth to philosophic reflection with a lucid detachment. His two wives, Saskia and Hendrickje are captured with great affection as models, as are his Jewish neighbours in group biblical scenes. He died impoverished, being both greatly generous and intemperate financially. It is hard to see Rembrandt except in his own huge perspective; his humanism shines from every canvas.

287 Jacques Louis David. (1748-1825) was a neo-classical painter working during the Revolution and the final years of the Ancien Regime. Influenced by Raphael and Roman mythology to paint heroic subjects such as the rather chilling 'Oath of the Horatii', he became an enthusiastic revolutionary, fostering, through his work, the heroic claims of the revolution. Briefly imprisoned at the end, he managed to survive and on release renewed his political enthusiasm, applying his heroic cast to Napoleon Bonaparte. His painting, *The Death of Marat*, is probably his most famous and chilling image. *The Oath of the Tennis Court* was commissioned but never finished for want of surviving models initially present.

288 Leni Riefenstahl (1902-2003 aged 101). The infamous film maker who made films for Hitler on the Nuremberg rally (*Triumph des Willens*-Triumph of the Will) and the 1936 Olympic Games with the famous record of Jesse Owens. Hailed as a gifted and innovative film maker in terms of technique, she always denied setting out to proselytize Nazism, although she frequently appeared with Josef Goebbels in German military uniform. Later she claimed she was simply naïve, and then that she had been threatened with internment. She was content to permit her fellow Jewish collaborators on her acclaimed film *Das Blaue Licht* to be removed from the credits in 1938.

289 The 'budding rose' is taken from Wordsworth's poem, 'The French Revolution', as it appeared to his initial enthusiasm…the most quoted lines are 'Bliss was it in that dawn to be alive/ But to be young was very heaven.'

Footnotes to Canto the Seventh

290   J.M.W. Turner's dates (1775-1851) are almost exactly contemporaneous with William Wordsworth. Turner came from modest social beginnings (his father and mother were both from trading families, wigs and butchery, which might have contributed to his sound head for business). A small, almost midget man, he was perhaps also dwarfed in other ways, except for his towering talent. He lived with his father (who acted as his studio assistant to the end of his own life), but was widely recognised in his own time, setting up both studio and gallery in Chelsea and building a fairly grand house in Richmond. He maintained loyal attachments to two older women, both widows, with one of which he fathered two children.

On his death he bequeathed a considerable sum and a collection of his works to the nation, never fully honoured, for his hope was to have his works collectively and permanently exhibited in London. Instead his legacy is probably best known for the prestigious Turner Prize, its sensationalism and vacuity a travesty of his influence, since his early works rivalled those of the Dutch sea painters while his late masterpieces were a major influence on the French impressionists. What a landscape of art he covered, from his first exhibit at the Royal Academy at the age of fifteen in watercolour, to his mastery of oil and its expression of his personal vision. His phenomenally adventurous late works moved towards the dissolution of the image, simply a celebration of light, enhanced by, rather than enhancing form. 'The sun is God' he is alleged to have spoken as his last words (in evocation of Parmenides). Nobody could better contrast the self-interested cynicism of David. The battle of Waterloo was fought when David was sixty seven and Turner was forty. What a world apart these artists were, less than 30 years difference between them in age.

291   The Czárdás is a traditional Hungarian dance, starting slowly and accelerating to frantic gypsy pace. The most famous was written for violin and piano by Vitorrio Monti but Czárdás have been incorporated in compositions by Brahms, Tchaikovsky, Sarasate and Franz Liszt. The term 'alla Polacca' in classical composition denotes 'in the manner of a Polonaise' as in Beethoven's Rondo in his triple concerto. The polonaise has a characteristic ¾ rhythm. A Mazurka is also Polish in origin, a lively dance with the emphasis on the second of three beats. The Dumka is a Ukrainian melancholy ballad.

Involution

292   The Baroque Sarabande is a very slow dance in triple time, derived from the much faster Spanish version. All the Bach cello suites have one, and the contrast they afford (after the running Courante) provided the grave measure retained in later string quartets from Haydn to Beethoven.

293   Johann Sebastian Bach (1685-1750). Although highly respected as a virtuoso organist, his compositional skills were not really recognised during his lifetime, until celebrated by Mozart and Beethoven. His understanding of counterpoint and harmonic progression were laid out in his monumental Forty Eight Preludes and Fugues written in every key. His compositions in Arnstadt and Leipzig were written for religious services, (including his monumental Passions for Easter) but his sojourn in the employ of Prince Leopold in Köthen gave him the freedom and the orchestra to compose his secular works: his orchestral suites, cello suites, partitas for violin and the Brandenburg Concertos.

294   Percheron is a breed of draft horse originating in the Perche valley in France. Intelligent and willing to work they are widely used for carts, timber moving and cross bred as war horses. They are usually grey or black.

295   Joseph Haydn (1732-1809) was probably the most prolific classical composer who, after an impoverished early life, enjoyed unusual patronage from Prince Nikolaus Esterházy for whom he acted as composer, orchestral organiser, chamber musician, and opera producer for over 30 years. Now called the Father of the symphony and the string quartet he was largely self taught in all areas, as Esterházy was a remote estate on the Austro-Hungarian border. Although unhappily married and without children, Haydn was highly regarded for his probity, good nature and extreme generosity. He was a good friend to Mozart who he praised to anyone who would listen. After the death of Prince Nikolaus he was retired and travelled to England, enjoying great acclaim as both performer and composer and writing the London Symphonies. The publication of these greatly increased his income, enabling him to settle in Vienna where, finally, he had the luxury of time. The results were his two great oratorios, The Creation and The Seasons. In all he wrote over 60 string quartets, 100 Symphonies, and some 30 concertos for a range of instruments. He made the piano trio

very much his own creation. His operas were less successful and are seldom performed.

296   Wolfgang Amadeus Mozart (1756-1791). His first six quartets (called the Haydn Quartets) were dedicated to Haydn, who after seeing them, said he could never equal them. After almost 60, Haydn never wrote another. Mozart's life-long pursuit of employment through all the courts of Europe, his psychological imprisonment by his feared father Leopold, his poverty and desperation and pauper's burial is one of the most tragic lives ever recorded. His one comfort perhaps was the brave loyalty of Constanze his wife who, as a trained singer, took a great interest in his work. After his death she astutely controlled the publication of his manuscripts and wrote the first biography in collaboration with her second husband, a Danish diplomat. 'Subito' means 'sudden', or 'immediately' in music. 'Da Capo' literally 'from the head' is a musical instruction to repeat a section from the beginning.

297   Ludwig von Beethoven (1770-1827). Unlike the painter David, Beethoven scratched out his dedication to Napoleon of the Eroica Symphony (No 3) with such vehemence that it left a hole in the paper, when he heard that Napoleon had accepted the title 'Emperor'. It was a denial of all enlightenment values he espoused, and a betrayal of all he had believed of him. His Variations on a Theme of Diabelli ran to 33, and instead of embellishing the almost banal simple theme with a single one, as requested for a charity collection, Beethoven wrote what is considered by some as the greatest composition for piano, or the 'greatest set of variations ever written' (Alfred Brendel).

'Sul ponticello' is an instruction to string players to bow as close as possible to the bridge which brings the hair of the bow close to the sound peg. This transmits not only the note but the friction, causing a rasping, breathy quality to the note.

Footnotes to Canto the Eighth

298   A Socratic dialogue is a dialectical method in which opposing points of view are pitted against one another with the objective of clarifying either or both.

299   Mount Kilimanjaro in Tanzania rises 5895 metres (over 19,000ft) above sea level, the highest peak in Africa. From most aspects it appears conical in shape but in fact has three volcanic cones. At one time the North Ice Field Glacier extended over 400 sq. kilometres but that has now shrunk and it is predicted it may become ice free as early as 2022. Kilimanjaro is visible from the Serengeti Plain and the Olduvai Gorge where man's earliest origins are thought to have occurred (Canto the third).

300   Beethoven's Grosse Fuge Quartet (Opus 130) ends with a highly complex fugue. This reference picks up where the last Canto ended.

301   Matthew 4:8 'Again the devil taketh him up into an exceeding high mountain, and sheweth him all the kingdoms of the world and the glory of them. (4:9) And saith unto him. "All these things will I give thee, if thou wilt fall down and worship me".

302   The following musical chronology, from the end of classicism to modernism, illustrates music's accompaniment to the dissolution of matter in the scientific understanding, as well as demonstrating that music equally reflects the collective. Its ground breaking geniuses step ahead, as do their scientific counterparts.

303   A title of a poem by John Keats (after a 15th Century poem of the same name).
    'Oh what doth ail thee knight at arms/ Alone and palely loitering/ The sedge has withered from the lake/ And no bird sings'.

Franz Schubert's romantic nature was expressed unbearably poignantly in many works. His friends teased him about his unrequited love for a pupil, Karoline Esterhazy, and Gustav Klimt painted a famous portrait of Schubert at the piano with four evanescent women standing behind him. He died listening to Beethoven's late quartet (Opus 131 in C sharp minor) and at his request was buried next to Beethoven in Vienna. Most of his work was unpublished in his tragically short life. He died at the age of 31.

304   Johannes Brahms (1833-1897) was born in Hamburg but worked and lived mostly in Vienna where he exemplified the Romantic developments in musical harmony, taking the classical structures of Mozart and Beethoven and extending them in new directions. His long devotion to Clara Schumann, to whom he dedicated many works and looked after following Robert Schumann's death, probably explained why he never married. Their letters to one another were destroyed.

Gustav Mahler (1860-1911). His work bridged the transition from the high romanticism to the beginning of modernism in the Second Viennese School of Schoenberg, Alban Berg and Webern. His composing output always took second place to conducting opera, particularly Wagner. His ten symphonies are all monumental works, apart from which he wrote only songs and one piano quartet. He was much admired by Shostakovich and Benjamin Britten. Largely self taught, (although encouraged by his father, a brewer and publican), Mahler performed his first piano recital at ten. He seems to have had very fractious relationships with orchestral musicians and colleagues and encountered great hostility from the anti-Semitic press in Vienna.

305   Brünnhilde was one of the Valkyries, and her creation was inspired by the Nibelungenlied. She appears in the last three of Wagner's Ring Cycle (*Die Walküre, Siegfried and Götterdämmerung*) playing a significant part in deception and fratricide and ultimately throwing herself on the funeral pyre of Siegfried.

Richard Wagner (1813-1883) began his career in the romantic tradition but later held a concept (much elaborated in his writings) that opera had to be a *'Gesamtkunstwerk'* – Total Work of Art – with as much attention to drama, poetry and staging as music. He developed the beginnings of chromaticism — the equal weight placed on each note of the chromatic scale — said to herald (in *Tristan*) the beginning of true modernism in music. This was possibly the legacy of his early interest in theatre and acting. He was particularly interested in Goethe and Shakespeare. Wagner wrote little besides his phenomenal operas but perhaps one of his most romantic gestures was his composition, the *Siegfried Idyll,* for his second wife Cosima, a transparently beautiful piece, to commemorate the birth of their son, played by a thirteen piece orchestra on Christmas morning–her birthday.

306   Edward Elgar (1857-1934) is, with Vaughan Williams, probably considered the quintessentially English composer. However he was more influenced by the Germans; Schumann, Brahms and Wagner, and in later life, when his music came to be recognised, held in higher regard by foreigners, especially Richard Strauss, Berlioz and Wagner. From the latter, as a self taught composer, he learnt much about orchestration. His *Enigma Variations* were musical portraits of his friends. This was a direct development of Wagner's leitmotif indications of 'characters' in his operas. Throughout his life he felt estranged and unappreciated, although his success in *The Dream of Gerontius* established a following in England. His violin concerto (commissioned by Fritz Kreisler) was quickly recognised for its genius; not so his Cello Concerto, although since its popularisation by Jacqueline du Pré, it is now accepted as one of the greatest concertos ever written.

307   Igor Stravinsky (1882-1971) was Russian born but naturalized French and later American. Unlike Elgar he was widely recognised and sponsored by a succession of wealthy donors, collaborating early with Diaghilev in his ground breaking ballets, *The Firebird, Petrushka* and *The Rite of Spring* written for the Ballets Russes. His influence across other artistic disciplines cannot be overstated. Unusually, after pushing the boundaries particularly rhythm and raw harmonies, he later reverted to a neoclassicism in such things as *Pulcinella* (in which he collaborated with Picasso). He developed motif composition which took themes from other composers and used them in entirely unexpected ways, often unrecognisable. His rhythmic invention which ignored the bar line continues to influence contemporary composers like Philip Glass. Stravinsky represents the starkest breakdown of the classical or formal musical language.

Footnotes to Canto the Eighth

Prokofiev (1891-1953) was another wunderkind whose talent was evident at five when he composed his first piano composition, followed at seven by his mastery of chess and at nine when he composed an opera. He began experimenting early with dissonant harmonies and unusual time signatures. Admitted to the St Petersburg Conservatory at eleven, he continued to defy conventional instruction and was considered precocious. Later in France he collaborated also with Diaghilev and in America with Stravinsky. After the war, and back in Russia, he was subject to strict Soviet controls and censorship. He died on the same day as Stalin, his passing hardly noticed. Some of his works such as his *Romeo and Juliet* ballet are widely known, others remain relatively neglected.

Benjamin Britten (1913-1976) was another early talent, composing from the age of five in various forms. His influence on others has not been as marked as those mentioned above. His easy facility was considered suspect in the English schools of composition but he was more highly regarded by his contemporaries in America where he was influenced by Stravinsky and Shostakovich. He is now regarded as a supreme opera composer of the 20th century (*Peter Grimes, Albert Herring, The Turn of the Screw, Billy Budd* etc) as well as a composer of song, mostly for his partner Peter Pears.

308   Arnold Schoenberg (1874-1951). Born in the Jewish ghetto in Vienna he was probably the most conspicuous revolutionary composer: moving from an initial romanticism which extended Brahms and Mahler (as in his *Gurre-Lieder Symphony*) before he turned to atonality using all twelve notes of the chromatic scale. He wrote widely on music theory and taught Webern and later John Cage. The equal use of the 'twelve tones' is perhaps music's dissolution into the un-patterned structure of thought itself (see Canto the Ninth) as it approaches the cone's apex and matter and energy are seen to flow, indistinguishable from one another.

Serialism is now the name given to twelve tone composition which was also adopted by Webern and Alban Berg, although Schoenberg claimed it was a natural progression from his earlier compositions and not revolutionary at all. After the rise of Hitler Schoenberg was unable to return to Germany and went instead to America where he was offered teaching posts in Boston and Los Angeles where he became friends with George Gershwin. He was also a very able expressionist

Involution

painter, exhibited alongside Kandinsky. Ravel was to say of his atonal music '*Non, ce n'est pas de la musique…c'est du laboratoire* (This isn't music, it's laboratory work).

The relevance of following the dissolution of musical formality and structure is to illustrate the artistic and non-verbal reflection of science's concepts of dissolving matter and the collective nature of consciousness.

309   Pointillism is the technique of applying small dots of pure colour (instead of blending colour on a palette) so that the eye derives the image from the contrasts between them. The artist Seurat developed this to its most sophisticated form. (Andy Warhol did so later.)

Cubism was developed by Pablo Picasso and Georges Braque as an avant-garde development from expressionism. The image is broken up and reassembled, portraying different aspects seen simultaneously on a single plane as though a composite image has itself been derived from different angles and perspectives. In this there seems an expression of relativity, wherein different views are all equally valid and none dominant.

310   Recent exhibitions at the Tate Modern have involved millions of painted seeds to create a space on which viewers were to walk (until there was a health hazard from chemicals). The 'Bed' is Tracey Emin's Turner Prize exhibit of an unmade bed, with cigarette butts, condoms, stained clothing. Another award was given to an empty room in which electric lighting came on and off seemingly at random intervals. Modern art might perhaps be called 'Empty Sensationalism' or the 'Moral Vacuum' in keeping with contemporary scientific concepts.

311   The '*disappearing Jester*' refers to Banksy, an anonymous Bristol born street artist who has now achieved world wide fame with his rapidly executed (usually at night) 'graffiti' paintings mocking capitalism and in fact any 'ism'. He paints on any surface he fancies, using spray paint and stencils, mocking images exposing inherent hypocrisy and double standards. His work is now achieving the prices hitherto reserved for 'great' classical paintings. He is very skilled at an economical marriage between environmental happenstance and an idea or comment upon it, such as two children using a 'No Ball Games' sign as a ball– far more dangerous, an appropriate comment on needless prohibition.

Footnotes to Canto the Eighth

312   While this Canto was being written, a report in *The Times* suggested that the Large Hadron Collider has apparently 'found' the Higgs Boson, the particle called the 'God' particle which would reconcile the disparities observed in the quantum universe, and explain the asymmetry between matter and anti-matter. The Higgs 'field' is a hypothesis that could account for the method whereby the non-material becomes material, accounting for mass in every physical manifestation…Plato into Aristotle perhaps? Is this the beginning of recognising that collective mind and matter have always been intimately unified?

313   In pre-Islamic Java the creation myth centred on Antaboga, the world serpent, from which was created the world turtle, Bedwang, and all subsequent creation was derived. Ken Wilber quotes a joke referring to this. 'A King goes to a Wiseperson and asks how it is that the world doesn't fall down. The Wiseperson replies "The Earth is resting on a lion" 'On what then is the lion resting?' "The lion is resting on an elephant" 'On what is the elephant resting?' "The elephant is resting on a turtle." 'On what is…' "You can stop right there your Majesty. It's turtles all the way down."

314   Strip-the-willow is a country or barn dance in which a line of men face their partners in a line of women. The 'top couple' at the head of the line swing one another before swinging down the line in succession, ending up at the tail of the line. The couple therefore 'samples' all the 'leaves' of the opposite sex. This seems an appropriate analogy for the progress of scientific reasoning; the thesis, then the antithesis, then the swinging together of both in the central synthesis, then the new derived thesis, taking in all the elements in succession.

Quarks are described by 'strangeness' and 'spin' values which distinguish them from one another, but do not mean what they seem to suggest to the non physicist.

315   Bose is an Indian physicist after whom the boson is named. He collaborated with Einstein in the Bose-Einstein quantum statistics to describe systems of identical particles to which Pauli's exclusion principle (that no two identical fermions can occupy the same quantum state simultaneously) does not apply. The particle now called a boson has a symmetric wave function which can be described by this

Involution

statistical system.

The boson was also the shortened form of boatswain (although spelled bosun) the petty officer responsible for a ship's rigging, decks and the duty summons to the crew.

316 A condensation of the history of ideas about matter from the Greeks to modern developments.

Magnetism was originally believed to have magical origins… hence 'Deceiver'. This dialogue pits the speakers on different threads of scientific discovery (on the left Magnetism, on the right Electricity) which, over time, will be united in small syntheses between them (in the centre). This expanded example serves to illustrate the process whereby opposing (and initially separate) disciplines move towards a deeper underlying law…thereby illustrating involution's recovery of earlier binding interactions of increasing uniformity. The acceleration of ideas towards uniform law is pronounced in this chronology, centuries separated the early ones, decades or years separated the later.

Theophrastus (372-287BC) a pupil of Aristotle (mentioned earlier for his botanical names) was the first to mention the property of attraction when amber is rubbed. Elektron is the Greek name for amber. So, at the origin of both, magnetism and electricity are bound in a casual observation.

Petrus Peregrinus, a friend of Roger Bacon, fifteen hundred years later (1269) mentions the attractive and repellent properties of the lodestone's poles, but likened it to the mariner's compass orientating towards the 'virtue' of the Polar Star.

William Gilbert (1546-1603) set forth the properties of the magnet in modern form. He predicted that at the poles (of earth) a free compass needle would point vertically; at the equator, horizontally. This he explained was due to the earth acting as a giant magnet or lodestone. In his book *'De Magnete'* he ascribes the effect as being 'animate' or 'spiritual'…'without error…quick…definite, constant… like an arm clasping round the attracted body…' He also perceived that iron was similarly responsive and as iron ore was mined in Magnesia so magnetism was named. In the opinion of Arthur Koestler this demonstrable action-at-a-distance paved the way for the acceptance of gravitation. It was Gilbert who coined the word 'electrics' to describe similar qualities in glass, sulphur, crystals and resins. He ascribed this characteristic to an 'electric effluvium' similar to magnetic effluvium.

Nothing was advanced in either field for a century until von Guericke, (1602-1686) the Burgomaster of Magdeburg, following work on atmospheric pressure, devised the first electric 'machine'. A globe of sulphur was made to rotate. Placing his hands against that rotation caused it to give off sparks, due to the electric charge imparted (by friction). Two similarly friction-charged spheres repelled one another.

Hawkesbee (in 1707) replaced the sulphur ball with glass, and later substituted the sphere with a cylinder. In the 1740's the glass electric machine had pads (rather than hands) and was connected to insulated copper wires which Grey and du Fey had shown carried 'current'. Current now replaced the vague 'effluvium'. Du Fey (1698-1739) had ascribed the attraction to one kind of current, the repulsion to a second kind of current; likening it to positive and negative magnetic poles.

Benjamin Franklin (1706-90) observed that charge could be 'drawn off' by metal points and ascribed this to a surplus or a deficiency of electric 'fire'. This was known as the 'one fluid theory' and he demonstrated the electric nature of lightning. Priestley suggested that the law of electrical attraction was the same as gravitational (calculated as the inverse square of the distance between the objects).

The first quantitative study was made by Charles Coulomb (1737-1806) using a torsion balance to measure the amount of torsion required to bring a charged ball within measured distances of another similarly charged, and therefore to overcome the repulsion. He thereby showed that electrical attraction and repulsion followed Newton's inverse square law of gravitational attraction. Cavendish studied electric eels and fish, and showed the mathematics devised by La Place for gravitation applied equally to both electrical and magnetic phenomena. (continued in note 317)

317   **The interchange-ability of energy.**
Poisson (1781-1840), the French mathematician, applied the equations of gravitational potential to electric and magnetic interactions by substituting 'electric charge' or 'magnetic pole strength' for 'gravitational mass'. This was the first striking unity of mathematical law underlying disparate phenomena.

It was J.J. Thompson (1856-1940) who suggested that the electron was the particle responsible for electricity, as well as being a fundamental particle of matter. Rutherford had already shown that the nucleus was positively charged. From this new and more dissected explanation for the behaviour of matter, in which it had been shown that an element's place in the periodic table was determined by the number of electrons in its atoms, Niels Bohr's atomic model was derived. The electrons circulated around the nucleus at fixed distances from it.

From this point all energy was conceived as interchangeable: Work into heat, mechanical motion into heat, heat into motion, motion into electricity, and electricity into motion or magnetism… conservation underpinned all exchanges from one form to another, one kind of energy to another. Joule (1843) embraced 'living force' as a similar kind of energy, 'it would be absurd to suppose it can ever be destroyed….whenever living force is *apparently* destroyed an exact equivalent of heat is restored'.

On a side alley a century earlier, Luigi Galvani (1737-98) an anatomist working on 'animal electricity' had performed an experiment of a dissected frog's leg hooked to a copper hook suspended on an iron railing. When the frog's leg touched the iron bar it jerked away. Since the iron was not electrically connected he concluded that the current originated in the muscle. It was left to Volta (1745-1827) to demonstrate that this conclusion was wrong: it was the contact between the copper and the iron that had generated the current (this being the prototype of the voltaic battery) with the circuit closed by the leg.

Later understanding would show that the muscle was responding to the nerve which was electrically sensitive to the closed circuit. So, electro chemistry was understood as also present in living tissue. The voltaic battery replaced the earlier Leyden jar, with its discharges of static electricity, by a flow of current with low potential but greater amperage (although those names would wait for Ampère). Humphrey Davy turned the basement of the Academy into an immense complex battery and paved the way for electric lighting.

Following work by Hans Oersted (1777-1851) which showed that a current flowing in a wire would cause the deflection of a needle in a compass, Ampère (1775-1836) made the brave supposition that all magnetic fields might be due to electricity; the one was the by-product of the other. To test this he placed a steel needle in the centre of a coil through which he passed a current. The needle became magnetized,

Footnotes to Canto the Eighth

and the first electro magnet was born.

So how did lodestones work, without current? Ampère suggested that minute currents were circulating in the atoms of the lodestone. These produced magnetic fields which aligned with the magnetic fields of the earth. This hypothesis was ignored for another century.

318   The genius of visualisation.

Michael Faraday (1791-1867), a great and versatile genius, realised that a current could be a source of power. A current creates a field of force (represented by concentric circles drawn around the wire). By an ingenious experiment he showed that by passing a current through a wire he could get a needle (or a magnet) to rotate, thereby transforming current into continuous movement…the principle of the dynamo. He then showed that a short lived current passed though a galvanometer whenever the opposite poles of two magnets were touched or parted (induction). Thus 'change' from one state to another caused electrical sparks and these were related to fields of force, the closer to the site of change the stronger the field. Faraday thereby introduced the concept of 'fields' to explain the inverse law. The greater the proximity of the opposing fields, the greater the force, the further away, the weaker.

Clark Maxwell (1831-79) extended Faraday's 'lines of force' by imagining the current carried in tubes, then eliminating (imaginatively) the walls of the tubes to achieve a gradation of forces within a continuous 'field'. From that, he concluded that all change (electrical, magnetic or hydrodynamic) sent 'waves' through space, with the same speed and character as light. 'We can scarcely avoid the inference….that light consists in the transverse undulations of the same medium which is the cause of electric and magnetic phenomena.' Electromagnetic radiation, having been now united, also encompasses light: both the visible spectrum and those extending beyond the senses, ultra violet, infra red, sonic, radio etc.

From now on science is concerned with the minute structure of matter to explain interactions. Maxwell did not go further to explain 'particles' of electricity.

319   The sub-atomic world.

J.J. Thompson in 1897 'discovered' the electron, which he proposed were present in all atoms. The 'indivisible, un-cutable' atom of the Greeks was now divisible. Thompson's was the 'plum pudding' model in which he visualised electrons evenly distributed throughout the atom.

In 1909 Rutherford bombarded gold foil with alpha particles and discovered that a few were deflected through greater angles than would occur with evenly distributed plums in the pudding. He concluded that the nucleus contained most of the mass due to the positively charged protons. This was the 'Rutherford model'. It needed further elaboration. Rutherford supposed that what he named neutrons in the nucleus provided the balance required. Neutrons were subsequently found by James Chadwick in 1932. Different isotopes of atoms were due to different numbers of neutrons. Stable atoms tended to have equal numbers of protons and neutrons.

In 1913 Niels Bohr anticipated the quantum nature of the atom by demonstrating that the electrons were in stable orbits around the positively charged nucleus. They could 'jump' from orbit to orbit but not to intermediate positions between them. These 'electron shells' explained the properties of the periodic table. This was the 'planetary model'.

Max Planck extended Bohr's understanding by showing that electrons were in 'packets' of energy which, when the electron jumped to a lower 'packet', resulted in the emission of photons, or required the absorption of photons to 'jump' to a higher one. These 'jumps' required fixed amounts of energy called 'quanta'. The Planck constant is the proportionality between the energy of a photon and the frequency of its electromagnetic wave. ($E = h\nu$ where h is the Planck constant.) This either/or nature of quantum physics is where the clash with classical physics lies, because 'a little bit more (or less)' is common to usual experience. However, in very small particle physics this 'granularity' of activity seems counter-intuitive to things we are supposed to conceive as 'waves'. De Broglie later showed that Planck's quantum described interactions between all sub atomic particles, not just electrons.

320 Waves or Particles? Uncertainty: the inter-relationship of mind and matter.

In 1926 Irwin Schrödinger proposed that all particles behaved like waves. Electrons were mathematically modelled as 3D 'waveforms' rather than 'points'. However the mathematics that resulted from this precise model found it impossible to obtain exact values for both position and momentum. This led to Heisenberg's Uncertainty

Principle which replaced the 'planetary model' with one of 'atomic orbital zones' governed by probabilities. The closer to nothing, the less exact the model. Mathematical probabilities govern the structure...or does the mind conceiving create the structure? Where does one begin and the other end?

In Summary: The atoms are now known to consist of smaller sub-atomic particles. In the nucleus are the protons and neutrons (called together nucleons) and composed of smaller particles still, called quarks. Quarks are bound together by the strong nuclear force mediated by gluons (one of the family of gauge bosons). The number of protons in a nucleus gives the atomic number of the element.

Different numbers of neutrons account for the isotopes (the variation within a single element) and the chemical bonding possible to form compounds. Atoms with roughly equal numbers of protons and neutrons are stable against radioactive decay, but the higher the atomic number the less stable they are.

At varying distances from the nucleus are the electrons in their 'orbitals'. These are the smallest particles and the closer they are to the nucleus the greater the force or heat velocity required to boost the escape of the electron from the strong nuclear force binding it. An electron can change to higher states by absorbing photons, or drop to a lower by emitting photons. Electrons furthest from the nucleus bond with other atoms to form molecules and compounds.

Very recent work by J. Hudson and his associates, at Imperial College London, seems to show that the electron, this most minute particle known, is the closest thing in nature to a perfect sphere. This has created new problems for the search for antimatter, which relied on some imperfection to account for the survival of matter, against a deficiency of antimatter. The positron (the anti-electron) which should have annihilated the electrons in equal quantities seems to have been defeated, perhaps by its perfection?

321 **Time and Time-reversal. The future determining the present?**

A positron can be regarded as an electron moving backwards in time. Maxwell's equations work equally well in both 'directions' with 'advanced waves' from the future interacting with waves from the past in the 'standing wave' that describes their interaction in the moment of 'now'. At the point of interaction all the probabilities collapse in the 'collapse of the wave function' to produce actuality. Involution as a hypothesis relates the continuous creation to such action in the 'now'. Consciousness moving backwards in time influences the changes to matter moving towards it from the past. In animals this 'apprehension' of the immediate future results in spontaneous behaviour that is encoded in the memory. In man that continues but is also recapitulated and augmented by the movement backwards in time collectively, as memory is progressively uncovered from the 'now' in early man, to the 'then' of his evolutionary past. Science has now recovered the all and is faced by the paradox of the origins: finite or infinite, mind or matter, time or eternity?

Nuclear fusion (as happens in the sun and stars) is the joining together of atoms to form heavier ones, giving off gamma or beta particles according to $E=mc^2$, a self perpetuating process once initiated. Fission splits atoms through radioactive decay which changes the atoms to different elements.

The photon, the charge-less, mass-less particle of light, is the fundamental gauge boson in which all properties are assumed to be zero. A photon is the quantum of electro-magnetic radiation and the carrier of electromagnetic force. Whenever a particle collides with its anti-particle it gives off two photons which conserve the required symmetry, each being the 'anti' to the other. In quantum physics they are considered 'entangled'. **Whatever happens to one will also happen (in the opposite mode) to the other, regardless of the space-time that may separate them. This entanglement is taken further in Involution, inasmuch as it posits the interconnectedness of everything throughout the universe, whatever happens in one 'locality' happens everywhere and forever, throughout time, as the field influencing every interaction.**

'Particle physics will breed new gnats,/Though filters like mosquito mesh/Will keep the little blighters out'. The 'filters' refers to the use of mathematics which maintains symmetry in the external world and prevents science from perceiving that the conjectures relating to particle physics relate equally to the processes of mind or consciousness (the little blighters). Rene Thom clarifies in his work *Mathematical Models of Morphogenesis* that the explanatory power of mathematics declines rapidly as systems become more complex. The need to describe the world in mathematical models probably contributes to the failure of science to tackle the question of consciousness, un-amenable to quantitative or statistical reduction.

322   The innocent questions and the Zeitgeist.

Albert Einstein (1879-1955), when asked about his 'discovery' of relativity, explained 'The normal adult never bothers his head about space-time problems...I, on the contrary, developed so slowly that I only began to wonder about space and time when I was already grown up. In consequence I probed deeper...' Minkovsky (Einstein's teacher) was quoted as saying 'In his student days Einstein had been a lazy dog. He never bothered about mathematics at all...'

In discussing Einstein as almost the quintessential genius in his naïve re-approach to asking questions, Arthur Koestler points out that his Principle of Relativity was 'unaided by any observation that had not been available for at least fifty years'. Poincaré, a much more sophisticated mathematician, had 'held all the loose threads but never tied them together'. With regard to space and time 'The spelling of the two words remained the same but their composite now signified something quite different from what they had signified before. Quantum physics has made the traditional meanings of words, like matter, energy, cause and effect, evaporate into thin air.'   (The Act of Creation pp183)

'When I use a word, Humpty Dumpty said in a rather scornful tone, 'it means just what I choose it to mean – neither more nor less' (Through the Looking Glass ch.6. Lewis Carroll)

"The most beautiful emotion we can experience is the mystical. It is the sower of all true art and science. He to whom this experience is a stranger ... is as good as dead. To know that what is impenetrable to us really exists, manifesting itself as the highest wisdom and the most radiant beauty, which our dull faculties can comprehend only in their most primitive forms - this knowledge, this feeling, is at the centre of true religiousness. In this sense, and in this sense only, I belong to the ranks of devoutly religious men." (Albert Einstein. In Marghanita Laski's Ecstasy 1961)

This highly condensed verse on Einstein wraps up a great deal that may require expansion, both about Einstein's revolutionary contribution to almost every part of physics, but also about the context of his genius. Other exemplars of such a sense of the numinous are discussed in the Appendix.

The famous Michelson Morley experiment in 1887 which seemed to show the speed of light as the same whether it was approaching or receding from the earth convinced Einstein that the idea of the earth moving through space had to be dropped. Newton's frames of reference were gone. All motion was relative: Time depended on velocity: Matter contracted in the direction of motion.

In fact, later repetition of the Michelson-Morley experiments *did* show some difference in the speeds of approaching and receding light but by that time relativity had entered the collective conscious and these later findings did not change that. D.C. Miller spent a lifetime trying to disprove relativity and his work seemed to do so, but Einstein had caught the cusp of the wave and Miller was too late to break its fall. Einstein said when the facts challenged the intuition 'the facts were wrong'! Many so-called constants are now being re-calibrated as the absolute standards are shown to 'change'. Could it be that involution has brought collective science to the interface of creation where mind alters matter and constants become relative?

323 Special and General Theories of Relativity: the micro and the macro unification through the mind of Einstein.

The two theories of relativity between them covered the physics of the universe; the Special Theory in 1905 relating mainly to elementary particle interactions and the General Theory in 1916 to cosmology, quasars, pulsars and the prediction in the 1930's of black holes and gravitational relationships.

The Special Theory governing the physics of space-time had two main components: 1. The laws of physics are the same for all observers in uniform motion relative to one another. 2. The speed of light in a vacuum is the same for all observers regardless of relative motion or the motion of the source of the light.

The implications of these affect everything. Simultaneity for one is not so for another. Moving clocks tick slower than 'stationary' clocks. $E=mc^2$: Energy and mass are equivalent and transmutable. Maximum speed is finite, nothing (exoteric) moves faster than light. (Thought, the esoteric, does not enter these hypotheses, although passive 'observation' does.)

To this the General Theory added the concept of 'inertial motion'. An object in freefall does not move in response to gravitation, but 'inertial motion'. This seemed to contradict Mach's Principle (which, roughly approximated, states that local physical laws are determined by the large scale structure of the universe). Einstein therefore devised his field equations for curved space-time which influenced the movement

Footnotes to Canto the Eighth

of mass within it and whose curvature was reciprocally influenced. The implications of the General Theory were then observed: orbital precession; the bending of light in the presence of large mass; the 'dragging' of space-time around rotating masses and the expansion of the universe. This is easily visualised by imagining the effects of curved space causing the motion as a billiard ball on a tablecloth moves if you lift the cloth. The causation is the field of curvature and distortion is that space in which matter is distributed. Each reciprocally creates the other. The mass distorts the curvature; the curvature influences the movement of mass.

In 1905, the year of Einstein's extraordinary contributions, four papers were published. In the first (On the Production and Transformation of Light) he suggested that light (or energy) existed in discrete quanta. These 'particles' or packets were called photons later by Gilbert Lewis, but the suggestion that light was particulate was rejected both by Planck and Bohr. However it was accepted in 1919 (after experimental verification on photoelectric effects and scattering) and could be said to have been the start of quantum theory which Einstein later rejected. His second paper (On Brownian motion) proved the atomic nature of matter and paved the way for the mathematical tool for quantum events, statistical analysis.

During the latter part of his life Einstein looked for a cosmological constant that would reconcile relativity that predicted an expanding (or contracting) universe with an unchanging one, which would more closely accord with Mach's Principle, but in the face of evidence for an expanding universe he admitted this was a blunder. He increasingly opposed the Copenhagen Interpretation of Bohr and Heisenberg by pointing out inherent contradictions relating to measurement. The 'wave function' is a mathematical calculation giving the various probabilities of the 'eigenvalues' pertaining to a particle. The moment one of these eigenvalues (spin, momentum, position) is determined, all the others 'collapse'–rather as a winning horse passing the finish eliminates any other winner. This in itself creates an asymmetric situation. The 'Many Worlds' theory gets round this asymmetry by suggesting that all the other probable eigenvalues exist but in 'other worlds'. This inherent contradiction lies at the heart of quantum theory which has to maintain symmetry or explain conservation in an asymmetric universe.

324   The Uncertainty Principle is often misstated so as to imply that simultaneous measurements of both the position and momentum cannot be made.

Published by Werner Heisenberg (1927), the principle implies that it is impossible to simultaneously both *measure* the present position while 'determining' the future momentum of an electron or any other particle with an arbitrary degree of accuracy and certainty. This is not a statement about researchers' ability to measure one quantity while determining the other quantity. Rather, it is a statement about the laws of physics. That is, a system cannot be *defined* by simultaneously measuring one value while determining the future value of entangled pairs.

325   **Quantum and Universal Entanglement.**

The Einstein-Podolsky-Rosen Paradox. (EPR). When two particles are created in a single event, the laws of conservation assume that each is the opposite of the other in values. If one has a certain 'spin' the other has the opposite and so on. The moment the eigenvalue of one is known, the value in the other is also known, instantly. This implies that what happens in one galaxy changes instantly what happens in another. The instantaneity is the problem, because according to general relativity nothing can travel faster than light. So how does one communicate with its mirror? Or just 'know'? The Copenhagen explanation is that the introduction of an 'observer' confuses, he is only the 'registerer' of what happens, and does not influence events. More recent analysis suggests that the problem lies in applying the laws of classical physics (i.e. the act of measurement) to quantum events. Quantum cosmology, to explain the creation of the Universe is now in a new infancy. The recent reappraisal travels in the other direction: quantum entanglement and quantum probabilities are now understood to govern macro events, from stock markets to tsunamis. It is further evidence for the collective penetration or involution to the origins of matter. Yet again underlying laws are binding irreconcilables together.

Recent theory has now recovered the concept of a cosmological constant as the Akashic Field (Ervin Laszlo), a deep Implicate Order of information (David Bohm), or universal memory (as here). Information present in this 'universal memory' (through the past coherence, or 'entanglement' of interactions between waves) would explain the coherence and integration of all created forms as well as the seemingly focused and accelerating nature of evolution. Science has now approached the integration of Platonic 'ideas' with Aristotle's evolution of form, as well as the reunification of God (as consciousness) and the Creation (as 'enfolded' mind, both single and universal). The language of DNA will be clarified in what follows, and its translation of 'mind' into 'matter' instantly and 'throughout' might reconcile this age long dichotomy.

326 The Large Hadron Collider (LHC) is the largest and highest-energy particle accelerator. It lies in a tunnel 27 kilometres long and 175 metres below the earth's surface beneath the Franco-Swiss border. It causes the collisions of opposing beams of particles, protons or the nuclei of lead, at very high velocities. The hope is to illuminate which of the opposing concepts is closest to an explanation of the formation of the universe. (The Standard Model or the Higgs-less model?) The Higgs boson is thought to generate elementary particles by breaking symmetry. Some of the questions it seeks to answer: Is super symmetry realised? (I.e. Do all known particles have super-symmetric partners, which would reconcile the Standard Model by contributing the quantum corrections and unifying three of the known forces?) They have yet to be found, but the desire for the corroboration of the model which has dominated quantum physics, would be grateful for their appearance.

 Are the other dimensions predicted by string theory detectable? What is the nature of dark matter that is needed to account for three quarters of the mass of the universe? (Could it perhaps be the unrecognised and unmeasured 'field' of consciousness?)

327 The 'hatchets' refer to the battle between opposites that has characterised the journey of thought: theses-antitheses, all integrated by syntheses. The journey now is to be beyond reason's oscillation, into transcendent experience.

328 'Bell, book and candle shall not drive me back,
 When gold and silver becks me to come on.' (Shakespeare. King John. Act 3 Sc3)

329 **Mind rediscovers the inter-connectedness of all phenomena: Recent Developments**.

The recent spate of books on new developments in science indicates that involution has almost run its course. Man collectively has re-penetrated the hourglass of memory to the origins of creation, and the limits of separate intellect. The Whole which pre-occupied the Greeks is now recovered in the patterns running through the whole, and the resurrection of the mathematics that pre-occupied them. A few 'maverick' thinkers perceive this in different explorations and analyses, now aided by the power of the computer to analyse vast quantities of data. What the Greeks apprehended instinctively, modern analysts need to derive from the model man has made of his mind- the computer, the tool that finally attests to the recovery of the journey. Perhaps the final return to the mind itself is implicit. (Perhaps too the prevalence of mind altering drugs, worldwide, is a deviant form of this search, beyond the material. Forcing the pearly gates and bringing man into incoherence?)

'The symmetries of water drops...' In 1994 Masaru Emoto froze water droplets from different sources and photographed their structure under a microscope. He found that pure mountain-streams produced perfect geometric crystals of hexagrammic shapes similar to snowflakes; polluted urban water produced misshapen distorted shapes. Further, by playing music to the water before freezing it, he found harmonious classical or folk tunes produced perfect symmetrical crystals; heavy metal rock produced distorted asymmetry. His conclusions were that only resonance to the music's vibrations could account for this. He is not, by many, considered a scientist and has been discredited largely because of his money making ventures on selling enhanced water. He is included here because belief is increasingly shown to influence matter, and at this point collective belief (through resonance?) may well convert once dubious claims to truth. (In the footnotes to Canto the Ninth recent work on the influence of sound waves on DNA and DNA's influence on laser light is elaborated.)

Emoto's work echoed the vision of Viktor Schauberger whose explorations into the strata and vortices of water made it, for him, a living pulsing substance. Water makes up two thirds of the biosphere and the human body, with the capacity to generate vortices from its tri-partite structure (a fundamental trinity of atoms, two oxygen one hydrogen) and a geometry that provides the medium to catalyse chemical reactions (and prevent others), transmit electrical currents, self clean and heal. (See Alick Bartlomew)

'*The death toll conflagrations/ The civilisations that collapsed*'. Since Einstein, science has equated energy and matter as manifestations of waves—from high frequency x rays to low frequency radio waves. Conversions, from one kind to another, power all modern technology, as well as technology's conversions to sensory input through the electro-chemical neurotransmission of the human nervous system.

Jean Fourier in the 18th century developed the mathematics to describe wave patterns which were recently used by Dennis Gabor in developing the investigations and theoretical framework of the hologram. Interference between waves (in which they augment through 'coherence'- uniform wave length and frequency as in lasers- or interfere through being dissimilar 'decoherence' and thereby producing different wave patterns) are the basis for achieving the hologram.

When coherent (laser) light is split in two, and one beam is reflected off an object, while the other is aligned to intercept the reflected beam (returning from the object) the resultant interference pattern is captured on a photographic plate. If this is then exposed to the original coherent light a 3D image of the object appears. Not only is the whole image visible in 3D but any small fragment of it contains the image of the whole. This is what is termed 'fractal geometry', the building blocks of the whole *is* the whole! (The One in the Many) Fractal patterns are increasingly found to lie at the fundamental structure of the Universe.

The quantum 'jumps' were originally thought to appertain only to the virtual world of sub-atomic event. Macro events, human affairs were believed untouched by it: No longer.

330 **Is Chaos chaotic?**

In the 1960's Benoit Mandelbrot studied a wide range of phenomena in human affairs: stock market crashes, weather patterns, erosion and volcanoes. These apparently chaotic and random events of systems in critical balance followed similar harmonic relationships, which he (mis)called 'Chaos Theory'. Such events were expected to follow the bell curve of distribution, the average clustered somewhere between the tapering extremes. Instead, he found a linear relationship: in the case of volcanoes the doubling of the Richter energy released reduced the likelihood or probability by four. This invariant consistent pattern is called a power law. Although you may be able to predict the likelihood of a given eruption, you cannot predict the size of the next. Again the law of the microscopic is found to extend throughout. Similar relationships have been found in the analyses of war casualties, and probably account for the cyclicity of civilisations in critical balance with the resources on which they depend.

One consequence of this fractal patterning is that a very small alteration (the butterfly wing) can cause either small or enormous consequences, it cannot be known in advance. All non-linear systems appear to obey this kind of power law: epidemics, the collapse of ecosystems, economic crashes, forest fires, tsunamis.

Holographic storage is multi-dimensional: not only does each part contain the whole, but if images are super-imposed upon one another, each contains the all-of-many like a palimpsest. Karl Pribram originally suggested that this may be the mechanism of brain storage of memory, which is why removing large sections of the brain often leaves memory more or less intact. Since the web of interconnected waves spreading out from nerve conduction would set up interference patterns, could the capacity of every cell to store the all be the basis of both memory and shared mind?

331 *Small is Beautiful.* The title of a book by E.F. Schumacher, which was widely read and influential in the sixties' idealism and its confrontation with big business, and global greed. The author was very kind to this poet in supporting a hypothesis that to science at the time was anathema and in some cases the stimulus to diatribe.

332 **Antiquity and Double Meanings.**
The Vedas (in Sanskrit 'knowledge' or 'wisdom') are the oldest Hindu texts (1500-1000BC), believed by some to be 'revealed' of other-worldly origin: roughly Bronze and Iron Age texts, conveyed by oral traditions in Hindu monastic settings. The oldest, the Samhita, are collections of metric texts providing holy mantras for recitation, other prose texts discuss and describe ritual (Brahamanas). When Sri Aurobindo first discovered the Rig Veda in Sanskrit he recognised its esoteric significance. *'A constant vein of the richest gold of thought and spiritual experience…I found that the mantras of the Veda illuminated with a clear and exact light psychological experiences of my own for which*

I had found no sufficient explanation either in European psychology or in the teachings of Yoga or the Vedanta.'

Sanskrit roots lend themselves to double or triple meaning. One can read these hymns on two or three levels of superimposed meaning…*a symbolism that is at least bizarre if one does not have the experiential key to the 'Fire in Matter'…the 'mountain pregnant with the supreme birth'*. This double or triple meaning in the earliest written language returns us to the 'language twisting twisting' of the shamans in the Amazon talking to the serpents, (Note 349) and which needs poetic allusion to suggest the multilayered understanding. (As well as the poetic recitation of Parmenides' visit to the seat of Apollo with the 'daughters of the sun' described in Canto the Fourth.) After the language of DNA is more fully explored in the next Canto the full circle of the evolutionary journey back to the earliest multilayered transcriptions will be seen as the most significant evidence of collective return to our origins.

After the end of the Vedic period the Upanishads and Sutra literature concentrated on philosophical and contemplative ways of attaining knowledge of Brahman. It falls away from its experiential base to an intellectual or conceptual one.

333 Rumi (1207-1273) was born in Rum (once ruled by Rome), and later his family moved to Anatolia, now in Turkey. Following his death and burial in Konya, his followers established the Sufi Order, the Mevlevi, known as the Order of the Whirling Dervishes. Through the influence of Shams e Tabrizi Rumi had turned from his work as an Islamic jurist to an ascetic; writing poetry about the union of Lover and Beloved (tawhīd) as the goal of spiritual life. His teaching acknowledged the equality of many paths to God which perhaps explains his considerable Western following. His beautiful writing in the new Persian language made his poetry widely read across the Persian Islamic world, and widely translated. Its appeal increases through its adaptability to various settings; music, poetic recitation and dance, celebrated particularly in Iran.

Footnotes to Canto the Ninth

334 The universal Serpent, its language and influence.

This summons of the serpent is to invite a re-evaluation of the entirety of assumptions about the nature of the universe and man's relationship with it. We have traced involution from the beginning of time, and through the history of science seen that 'believing is seeing'. The concepts of perfection shaped the enquiry, whether in form (perfect circles, regular motion) or the destiny. After the Scientific Revolution in the 17th century, destiny was eliminated in the mechanistic material universe that alone was the proper study of science. This I hope to have seriously upturned; by showing that involution— the incremental recovery of memory– has always governed the sequencing of scientific modelling and the progress of mankind. Involution has taken us to the gate of creation, to the origins of both matter and mind.

This 'personification' of a speaking DNA may appear metaphorical artistic licence, but may now prove to be literal, inasmuch as DNA holds the historical records, and in its language speaks creation throughout the living universe. Recently, work in the Russian Academy by Pyotr Garjajev and colleagues who have approached DNA as linguists not as chemists, has now uncovered the archetypal first language in DNA. This is clarified below, after the structure of DNA is offered in some detail.

Despite the recent suggestions of the need for 'Morphic Fields' or the 'Akashic Record' DNA is not, with a few recent exceptions outlined below, considered a likely source of mind, perhaps because the established link with brain is too entrenched, yet bodily and cellular mind has been demonstrated in the lowest creatures (see Canto the First) and long known in the mystical traditions. Yoga, in various ways, seeks to overcome the autonomy of bodily 'mind' by higher forms of consciousness. What this soliloquy seeks to emphasise is both the universal spirit in all, and its uniqueness in each (held within the all); in other words the One, the Many and the All. All perennial philosophy claims that the hierarchies of consciousness are ascended through the abeyance of the ego-will, and its intellectual constructs. This loving address by the serpent simply repeats that necessity as the next phase of Involution/ Evolution, the realisation of cosmic consciousness beyond intellect.

335   Caduceus. In Greek it literally means 'herald's staff' and was traditionally shown carried by Hermes. It shows the staff entwined by two snakes, often mounted by wings atop. In Roman iconography it was carried by Mercury (and came thereby to be associated with the elemental metal of the same name). By extension it was associated not only with truth and eloquence, but trade and commerce. Hayes Ward found it reproduced on much earlier Mesopotamian seals, putting its origin somewhere between 3000 and 4000 BC. The rod of Aesculapius (with which it is commonly confused) has only a single snake.

336   The Medlar fruit was cultivated by both Greeks and Romans from the 2nd century BC. It is native to Asia Minor and the Black Sea coast of modern Turkey. The Mespilus tree (which bears the fruit) is of the Rosaceae family, so the Medlar 'garden' carries the connotation of early Eden and the first 'apple' as well as the rose.

337   Frangipani is a flower of the Plumeria genus, native to Mexico, related to the Oleander. The flowers are headily fragrant, particularly at night when their scent dupes the Sphinx moth into pollinating them, without offering any reward in the form of nectar, which they do not have. The trees are associated both with death and worship, and are commonly planted in graveyards or in temples in the Philippines and Bali. It is known there in English as the 'Temple Tree'.

338   This refers to the statement of Newton. 'If I have seen further…it is by standing on the shoulders of giants'. (In a letter to Robert Hooke 5th Feb. 1676 quoted in *Correspondence of Isaac Newton Vol. 1* ed. H.W.Turnbull. 1959).

339   This relativity is the destination suggested for consciousness, and perception. As space and time became space-time, so as intellect and consciousness merge, the perception of the 'otherness' of the material world becomes relative. Consciousness is a quantum phenomenon.

The more intellect dissolves into wider consciousness the less 'material' or 'solid' is the perception of everything; matter and energy are perceived as two aspects of the same thing, just as particles and waves are both characteristics of microscopic matter. The unification of previous disparities has been the pattern of science throughout. Mystics, throughout time, have described the dissolution of the solid world into the white light of 'beingness', the 'uncreated plenum', the 'fullness of nothingness'; all paradoxes and distinctions melt into unity in this experience.

340 **Visualisation, mathematical models and the incremental involution from Now back to the Beginning: DNA speaks and records throughout.**
Before proceeding to a description of the 'form' of DNA it is as well to recapture the many previous examples in which visualised geometry gave rise to scientific hypotheses about shape and motion. From the crystalline heavens of Aristotle, to the harmony of the spheres for Kepler, to the perfect orbital motions of the planets, to the planetary models of the atom, visualisation has always preceded understanding. These were all critical imaginative steps to a refining approach and as such useful for the period in which they remained; each giving way to better 'visualisations' or mathematical models. Science builds on what it already knows, and thereby creates (from the past into the present). Mystics experience first and then find ways to convey, from the perennial back to the language of the present. Teilhard de Chardin, both a scientist and religious visionary saw the 'field' of information in his 'Noosphere' but was sixty years ahead of the scientific climate of consent.

The poet here is suggesting that as measurements took the human body and its proportions as the starting point for analogous comprehension (see Cantos 1 and 2 and the construction of the temple of Luxor, or Leonardo da Vinci's Vitruvian man's circles and squares), so involution has brought mankind to the apex of the cone of creation. The model of DNA's structure might prove to have been, all along, the basis of mathematical modelling, both in its chemical structure and its causative creation of all living forms. All creation relies on a narrow choice of letters to speak a wealth of creative 'language'. The Fibonacci series' echoes are everywhere, and now (according to Penrose and Hameroff, see Footnotes 375, 376) may govern the connections in

Footnotes to Canto the Ninth

the micro-tubular lattices that enable cellular computation, allowing for a computational 'mind' even in non-nucleated cells. This model of computational ability is further expanded in the understanding of DNA's contextual structure, its rules and patterns, which now are understood as additional requirements, underpinning its 'creative' material function. It both speaks and creates.

341   **The geometry of DNA.**
Some understanding of DNA's chemical structure and habits is necessary to make sense of these analogies, but I hope to make it simple, so that the claims of the serpent are not dismissed as merely metaphorical. The original Watson - Crick Model essentially applies to the 5% of the genome responsible for protein manufacture. Once this is understood, what follows– the internet of DNA– will be better conceived as it is a language yet to be explored. It seems that our tools have preceded our ideas in recent times.

Some conceptual visualisation of the structure of the molecule will assist.

Size and Quantity
The size of a human body cell is almost exactly half way between that of an atom and the human body itself. Every cell in any living creature contains the same DNA, in quantity and structure. The language of DNA is universal, consisting of the same alphabet and grammar; creatures differ only in the quantity and context of that language present in their cells.

DNA is held in the nucleus of every cell, invisible even to the light microscope, except at certain phases in its duplication before cell division. As the control centre of the cell and through its instructions, DNA determines the organs and form of the body.

The cell is filled with water, almost identical with sea water, and the nuclear membrane confines the DNA. Access to and from DNA is through small holes in the nuclear membrane through which small molecules are permitted selective access to receive and convey instruction. (Rather as the secretaries to the CEO of a company permit limited access.)

The DNA is arranged in chromosomes which are present in two copies, always paired, except in egg and sperm before fertilisation which have one copy of each. (Fertilisation restores the full paired complement.) The longest human chromosome fully unwound would stretch 100 km (the distance from Cambridge to London) and altogether in the human body estimates suggest 125 billion miles of DNA. So the information is doubly double. The strands of DNA are mirror images of one another, the chromosomes are equally doubled.

342 **DNA Arrangement and Structure**
Different species differ in the number of chromosomes, always arranged in pairs plus the X (female) and Y (male) which determines sex in species where sex is separated. Humans have 46 (22 pairs plus XX– female or XY– male) the fruit fly has 8 very large and conspicuous chromosomes (3x2 + XX or XY) which is why most laboratory studies on genetics were originally understood through the fruit fly.

When the chromosomes are stretched out they show striated bands of dark and light (like a bar-code); these patterns are species specific, the dark bands consisting of much more protein mixed with the DNA than the light inter-bands. This mix of protein enables the DNA to be compacted: the protein arranges itself in spools around which the DNA wraps itself like thread, reducing its length by a factor of six but preserving the critical 'sequence' in sequence.

The form of the molecule of DNA is a double right-handed (clockwise) helix, but the two strands run in opposite directions, anti-parallel. This is a result of the mirror image structure of the very specific bonds between them, which both hold them together, and ensure the identical duplication. The units of the chain are called nucleotides, and each nucleotide consists of three elements, a sugar in a ring shape, a phosphate and a base. (Thus 'trinity' starts with each letter of the alphabet) ('*My symmetries of coil /my alphabet of two placed hands...*') but goes further as we shall see.

343 **The fine Structure of the DNA molecule, its alphabet, and how RNA 'reads' that to enact protein manufacture.**
There are four different bases, called Adenine (A) Guanine (G) Thymine (T) and Cytosine (C) and the ways in which these pair are totally specific: A pairs with G and T pairs with C, due to the arrangement of their atoms. This specificity accounts for the accurate duplication of the DNA molecule. When the two threads are separated as though 'unzipped', the exposed docking points attract only the base that fits the dock…A attracts G, T attracts C. The two strands are held

together by Hydrogen atoms, two between A and G, three between T and C. The change to Uracil (in RNA) gives much greater flexibility of pairings, and would lead to many more errors in copying. This is perhaps why DNA replaced RNA as the genetic code. A single change from G paired with C to G paired with T can cause cancer.

There are therefore only 64 letters in the DNA alphabet (4x4x4=64). How these convert to a language is through three letter words of adjacent bases (AAA, AGA, CAA, etc). The trinity continues in the structure of words (a biological terza rima). This code is converted into structure by the manufacture of proteins. The genetic code for protein manufacture is equally universal, although creatures differ in the proteins their structure requires. (*My language speaks in trinities…*)

Proteins consist of long chains of amino acids, of which there are approximately 20. By combining these 20 in specific sequence chains of 100 to 1000, different proteins are manufactured, some are structural like muscles, tendons, cartilage, some are enzymes catalyzing other chemical reactions, some, like haemoglobin, are carriers, transporting oxygen or other chemicals. Their structure is their function.

The sequence of the DNA codons (the three letter words) determines the sequences of amino acids that assemble to make each specific protein, in specific quantities. So for example AAA codes for the amino acid Lysine, the code TTT codes for Phenylalanine and GCT codes for Alanine. A few triplet codons code for more than one, and other codons are punctuation points; 'Start here, Stop here' 'Repeat this part'. This universality of language explains why genetic modification can be introduced…a sequence taken from one species will be read by another and a 'new' protein (new to the host) manufactured under new instruction from the universal amino acids present.

344   **How is Protein Manufactured? The translation from DNA to RNA and to Protein Synthesis.**

Protein is manufactured outside the nucleus. First the gene (a section of DNA) makes a copy of itself as RNA (this is almost identical to DNA but instead of Cytosine RNA has Uracil with one more Oxygen atom. The Uracil bonds with Thymine instead of Cytosine). This 'messenger RNA' leaves the nucleus, passes through the membrane to the Ribosome (the workbench) in the outer cell, carrying the instructions from 'headquarters, the nucleus'. (*Short injunctions are to matter sent…)*

In the outer cell 'transfer RNA' acts as gofer, ferrying the amino acids that will be needed in protein assembly to the ribosome. The ribosomal (workbench) RNA helps to align a series of transfer RNA molecules (carrying amino acids) with the messenger RNA to make the specific protein chain. Because RNA is critical to each step of the assembly it has long been thought that it was the original genetic storage and instruction mechanism–the first genes. This idea is beginning to change as its flexibility is now seen as crucial to its role as 'messenger' not only from the DNA (as in protein assembly) but back to it (as in constant modification).

A protein RNA polymerase (an enzyme) causes the DNA spiral to unwind in short sections, separating the strands for very brief seconds (100 base pairs, or 'rungs', a second) and thereby permitting messenger RNA to take the code from the unwound section before it coils up again to hide the structured information from dissolution in water. In practice the DNA spiral slides through the polymerase uncoiling a section at a time. The TATA (Thymine Adenine pairing) are sequences where the spiral most easily unwinds and probably promote the start of the 'reading', other short sequences are 'stop codons'; they stop unreliable readings that would result in copying errors.

345   **Specificity and Superfluity. What is understood is the 'coupling' of DNA and RNA to form proteins. If this is all DNA does , and only 5% at that, what purpose is served by the 'junk or 'silent' DNA? What is the purpose of its superfluity in every cell? What else might it code to explain the highly organised structure of this 'junk'?**

It is known that only 3-5% of DNA codes for protein manufacture. The remaining DNA is called 'junk' or 'silent'. Its purpose is unknown, conventionally disregarded as merely the legacy of evolution— the extinction of irrelevant information (or nature failing to empty the bin!). It is also superfluous in its vast repetition. Only the original sex cells, eggs and sperm, need a full set of instructions. After a certain stage in embryology when the original single cell has duplicated many times, specialised cells in different areas of the embryo form brain, liver, lungs, heart etc. Yet each and every cell retains the instructions for the whole. This is the holographic nature of nature, each is itself and each has the whole within itself. More about the significance of this follows.

The coiled shape of the DNA molecule is essential to maintain the stability of its structure. (What follows this conventional account suggests that the way in which the mirror strands align is also critical for the order in which the information is 'read'–see below.) The bases are insoluble in water, unless attached to phosphate or sugar (both of which are soluble). This potential solubility in the environment of the cell is solved by the spiral. The paired bases forming the 'rungs' of the ladder are tucked on the inside of the coil whose rotation places them one above the other, protected from the water, and the exact coiling necessary can be calculated to avoid 'holes' by rotating the strand. The coil does more than a simple 'twist' would. It not only eliminates the holes but holds the bases apart at critical angles to one another (32'3 degrees or one eleventh part of a complete circle or eleven phosphates per turn).

Although both clockwise and anticlockwise spirals in theory could exist and the Z form (with 12 phosphates per turn) is known, in practice almost all DNA is clockwise or right handed and almost all run with the two strands being anti-parallel, running in opposite directions. In brief; the exactness of DNA's geometry is its function for stability and accurate duplication. The back-to-front (mirror) nature of DNA's strands is important to the hypothesis of involution which runs in the opposite direction to evolution i.e. anticlockwise, – back through the records of time. (By a time-reversal chronology and as an anti-material component, consciousness, it also provides the complement to time and matter.)

346   Throughout this hypothesis of collective and chronological recovery of stored memory, the critical breakthroughs have been deemed reliant on the single 'genius' who saw deeper than the collective intellect of science. What characterised these 'Eureka' moments was the certainty that accompanied the insight, although as yet untested by measurement and tools. (Like messenger RNA they retrieved fragments of the pattern and took them 'outside'.)

347   **The geometry of DNA's duplication, its phases and orientations.**
The extraordinarily stable DNA (or its predecessor RNA which it has harnessed, as the cell harnessed bacterial mitochondria and chloroplasts) (see Canto 1 note 65) outlives anything else. Cells have

been found and dated 3.9 billion years old and DNA has changed everything, but in orthodox assumptions in evolutionary theory been itself unchanged-except by incremental errors. Bacteria do not have nuclear membranes but the algae that created enough oxygen to support life were the cells to which all owe a due.

The compaction of the DNA molecule (by associating with and wrapping round proteins) has been mentioned, but further compaction into the chromosome is due to the coiling up of the protein spools as well. Until the separate chromosomes take up a geometric formation (before cell division) the globular mass of chromosomes is difficult to distinguish. It is even at this level of the cell nucleus that trying to interfere with the structure (in order to understand it) threatens to destroy what it is under study. Less is known about chromosomal behaviour than DNA duplication; even the 'banding' is not fully explained.

What is better understood is the behaviour of chromosomes just before cell division. Somewhere in the centre of each chromosome is a structure called the centromere, (long semi-repetitive sequences of base pairs) and at each end other structures called telomeres, (short repetitive sequences like TTAGGG) repeated hundreds of times.

Before the cell is about to divide, protein 'micro tubules' assemble across the polarized axis of the cell. The centromere of each chromosome attaches to these tubules and they are pulled apart from one another (un-zipped), one chromosome of each pair moving away from the other, until separated by the division of the cell. The telomeres have uncertain purpose: perhaps sealing the ends or attaching to the nuclear membrane to anchor the chromosome in 3D space during cell division. (*'My pyramids are sometimes set /to conically divide...'*) It is interesting that polarity develops in the otherwise undifferentiated nucleus just before division and the creation of new fertilised life. It seems to anticipate the necessary 'opposition' within the material universe into which new life will enter.

348   This suggestion assumes the 'echoes' within DNA, which both holds the records of all, and in which each has a part. The 'in-commonness' throughout biological evolution of a structure which binds the history of all, and specificity of each, may provide the necessary resonance that accounts for the fine tuning of each to the all.

349 **Visualisations of DNA in primitive art, communications between mind and DNA serpents, double language; double helix**.

Further corroboration is provided by Jeremy Narby's book 'The Cosmic Serpent'. He gives persuasive evidence for the universal mythology of the 'serpent' across all time and many civilisations. The serpent is portrayed as sometimes single, sometimes twinned (as are DNA strands). The Peruvian shamans draw the serpent(s) in the different forms DNA adopts during cell division; lined up next to one another as in the prophase start to preparations for cell division. In the Anaphase the two 'snakes' adopt an X shape (connected at a single point) and the double conical polarity is visible before the sister chromosomes separate. After cell division each duplicates itself to restore the double snake for each of the new cells. Both the chromosomes and the DNA molecules go through this separation and reunification. The anti-parallel orientation of each strand of DNA is captured in both the double headed snake, and the two snakes with heads at opposite ends.

Narby has done a great deal of anthropological work with the Amazonian shamans (or Ayahuasqueros) who take an hallucinogenic drug, Ayahuasca, in order to communicate directly with the serpents, from whom they claim their source of wisdom regarding natural causes, medicinal drugs, and the unity of all life. (Ayahuasca is a powerful drug which contains a substance similar to serotonin, the neurotransmitter and Dimethyltryptamine (DMT) which is manufactured in the Pineal Gland, long believed the site of the 'thousand petalled lotus' at the crown of the head, where the Buddhists believed, as did Descartes, that the soul connected with, and escaped from the body.) These shamans communicate their vivid visions in an oblique language called 'Language twisting twisting' which uses poetic imagery to capture the multi-dimensional nature of what they perceive. In other words, the necessity for the language is to try and capture the structure of a double double process, of association and constant transformation and interconnection. Nothing is one thing; each is also everything. At this point experience recaptures Heraclitus, or each thing embodying and being transformed into its opposite.

350 The cortex is the multi-folded surface of the brain, the most recently developed, to which is attributed reason, analysis, and memory. The serpent here suggests that the purpose of the cortex is to receive, as a

Involution

radio does, the signals, from the cells of the body and from elsewhere, and through all language to convey them. See the language of DNA posited below.

351   **Quartz and Crystals. Amplified vibrations, and junk DNA's sequences: similarity to the periodic crystal, quartz. Bio-photon emissions and reception.**

Since time immemorial quartz has been associated with both shamanic practices and alchemy, frequently found in prehistoric religious sites, and widely used in healing practices. Quartz is a crystal with a very regular arrangement of atoms which vibrate at stable frequencies, amplifying vibrations with which it comes in contact. It is, in that sense, a sort of vibrational microscope.

DNA is also a crystal, an aperiodic crystal, due to the 'stacking' within the helix of the hexagonal and pentagonal bases, one above the other. However in the so called 'junk' sequences, where similar bases repeat in endless arrangement of ACACACAC…these form areas of DNA that are regular periodic crystals and therefore able to pick up and amplify vibration, just as the quartz does. This capacity of the 'junk' sequences of DNA is further understood and detailed below: its vibrational and retentive capacity for direct communication of all with all— these may be the 'strings' of recent quantum theory or the basis of fractal geometry and patterning.

It has recently been shown that all cells of all life emit a constant stream of bio-photons, and Narby and Popp have both suggested that DNA is the source of this constant emission. The frequencies of these emissions fall within the band of visible light. Rattemeyer et al (1981) originally suggested that the role of junk DNA sequences was to emit electromagnetic radiation. Narby suggests that given its crystal amplifying structure DNA could, through harmonic resonance, 'pick up' as well as 'emit' bio-photon vibration. This would explain both the coherence of an integrated universal 'mind' as well as the holographic nature of the universe following the same patterns throughout creation. Recent work in Russia has shown that his assumption was correct (as was Involution's supposed mechanism in 1970). More detailed understanding from Russia clarifies this phenomenal capacity below (see Note 383).

It also links the incoherent (for want of 'language twisting twisting') results of scientific inspiration. The Ayahuasca hallucinogen introduces shamans directly to the serpents of DNA, which speak directly to mind with linkages for which no language exists. They invent poetic metaphor; science does not honour or attend to metaphors!

352 **The relationship between the geometry of DNA and mathematical modelling?**
The language of western science has been mathematics, and the serpent here puts forward a claim that its form and structure has provided the basis of understanding: angles, triangles, the patterning of number, conic sections, and circular motion are all to be found in the chemical structure of DNA. *(My language is of molecules...)* Given 125 billion miles of DNA in the human body, the reception and emission of information through vibrational sensitivity within each cell approaches limitless, infinite dimensions. Canto the first began with man's measurements being modelled on his body (feet, inches, yards etc). It is a logical extension for intuitive men finding repeated patterns throughout the universe to intuit the mathematics of the molecules that connect it all together in the molecular mind. More plausible now that the universal language of DNA seems to prefigure all symbolic language, including mathematics (see footnote 383 below).

353 *The Light of the World* is a famous Victorian allegorical painting (now in the chapel of Keble College, Oxford) by William Holman Hunt. It shows Jesus knocking at a long over-grown un-opened doorway. 'Behold I stand at the door and knock...' As the door has no handle it can only be opened from the inside (the closed mind or heart?). The artist claimed he painted the work following divine inspiration.

354 **The survival of memory and circumstances of its recovery**.
The indestructibility of life (and the permanence of evolutionary memory) has, through a belief in re-incarnation, been held by many, if not most cultures. Stevenson, mostly ignored by science, spent his life collecting direct evidence of past life memories in young children (not only in the child who remembered, but in the extended families who recalled previous encounters with them, and could produce objects recalled (from a previous life) but never seen in the current one). To these we must add near-death experiences, and out-of-body

experiences. This wealth of evidence is now making survival of memory and re-incarnation more open to enquiry. Past life therapies take the legacies of past lives, often detrimental to mental health, as their starting point (see Roger Woolger). Only Western Science refuses to examine this widespread wealth of evidence. The Buddhist belief that the purpose of re-incarnation is to afford the soul a gradual refinement through the body to finally escape the 'wheel of life' has been pursued for millennia…and, in essence, is no different in message from that of Parmenides and Socrates. Involution has merely retraced us back to its necessity now, and its inevitability. Once open to its role throughout, mankind will perhaps be open to experiencing what mystics of all faiths have sought and lived by.

355   The Lorelei - literally 'the murmuring rock'. (From the old German *lureln-*to murmur and the Celtic *ley-*rock.) The Lorelei rock stands at the narrowest part of the Rhine where strong currents and invisible rocks have brought many men to drown. The high rock amplified the sounds of the currents and a waterfall, leading to the mythology of a beautiful siren woman who bewitched men to their ruin. Heine wrote a poem called Die Lore-lay which recreated Ovid's story of Echo.

356   **DNA altered by experience? Evidence for the susceptibility of DNA to alteration in vivo.**
'A Universe of spies'. When originally written (in 1970) 'Involution' relied upon an assumption that DNA was progressively altered by experience. At the time of emphatic Darwinism this was considered heretical (because it resurrected discredited Lamarkianism; at that time DNA was a dictator, subject only to 'error'). It is now being shown that DNA is altered by multiple 'switches' that respond to diet, mechanical stress and radiation. What is now called 'epigenetic' inheritance refers to inherited habits that do not require changes to DNA structure, so called because the assumption that genes are not changed except in error is inviolate.

Recent research by Andrew Maniotis showed an instant mutation in the gene caused by mechanical force applied to the cell membrane. Unlike most cell biologists he studies intact cells, and in particular what he calls tensegrity, the structural ability to maintain integrity through tensile structures, similar to those described below by Penrose. (Significantly, the different behaviour of DNA in vivo from its relative passivity in vitro conveys its dynamic and necessary interplay with its immediate environment.) Maniotis has demonstrated that DNA has structural as well as instructional importance, and in his words 'the structure is the message'; in short, a corroboration of the hypothesis underlying the process of involution. Michael Lieber has also demonstrated that any 'stress' coming from the interaction between cells and the environment causes global 'hyper-mutations'. His work has concentrated on mutagens, (substances or processes inducing mutations) and a critical evaluation of the limitations of the 'molecular reductionism' of genetic causation as generally conceived.

So the dynamic interplay between organism and the environment is registered in the genome constantly. Only such a dynamic could explain the rapid adaptation of creatures, and particularly man, to changing conditions, extreme temperatures, and new colonisation. The interconnectedness runs throughout. These recent developments increasingly support involution's suppositions with palpable processes. Now it is understood that DNA can respond directly to sounds and polarized photons, retaining the information of past encounters. The 'translation' of its coding into matter is increasingly understood as the most 'pedestrian' of DNA's functions, despite its perfection.(see Notes 358 and 382)

357 The Beagle was the ship on which Darwin sailed for a four year journey round the earth, and during which he perceived the great disparities of plant and animal life, and their specialisation when isolated by land masses or geographical barriers. From these observations he first formulated *The Origin of Species*. Deep time was the insight of James Hutton who realised the geomorphology of the earth indicated an age far greater then previously believed. Lyell's support of Hutton greatly influenced Darwin.

358 **Hyper-communication between DNA through light and sound.**

Recent work in Russia provides another dimension to this claim (in addition to the work of Penrose and Hameroff annotated in footnotes that follow) and widely elucidated in 'Vernetzte Intelligenz' the work of Fosar and Bludorf in Germany. They report on research by Pjotr Garjajev showing the alteration of DNA, *'Living chromosomes function like solitonic/ holographic computers using the endogenous DNA laser radiation.'* A laser ray, whose frequency patterns were modulated, influenced the frequencies of the DNA emission and the information itself. Living DNA (in vivo only, not in vitro) will react to 'language modulated laser rays' and even to modulated radio waves which are shown able to repair damage caused by irradiation, as well as able to transmit the information from one biological form (frog for example) to another (salamander). The use of sound in altering states of consciousness has long been known in the mantras of Buddhism.

The most exciting part of this work is the suggestion that DNA frequencies can cause 'magnetised wormholes' (the microscopic equivalents of the Einstein Podolsky Rosen (ERP) 'bridges' found near black holes) and through these 'tunnels' information can be transmitted outside space and time (i.e. faster than light). 'The DNA attracts this information and passes it to consciousness'.

This 'hyper-communication' occurs when the mind is 'switched off and relaxed'. (Supporting the examples annotated in this work on genius and extended in the Appendix.) They further suggest that early man and early civilisations had the ability, as do animals, to respond to this hypercommunication but that the ability has been lost in the separation of intellect,(the 'Exile' this book describes); thereby giving explicability to the 'holistic' genius of Egypt and early Greece. Swarming insects such as ants are shown to continue to build the colony even after the queen is removed, no matter to what distance, but they cease the moment she is killed. This ability for DNA to retain information even after the source of transmission is removed is called 'The phantom DNA Effect' and its recent discovery could explain the retention of evolutionary memory. I acknowledge this work is very new, and much remains to be understood. (More detail is provided in Footnotes 383 and 384 below.)

359   **The entanglement of micro and now macro events.**
Schrödinger's cat is a famous 'thought experiment' to illustrate the *reductio ad absurdum* of the Copenhagen interpretation of quantum entanglement and the principle of indeterminacy— that a quantum state is a combination of two (or more) mutually exclusive certainties—which collapses (since they are both probabilities) when one is established.

A cat is placed in a sealed box in which there is a phial of deadly poison, and a source of radioactive decay, which may or may not decay. If it does, a Geiger counter is set to respond by breaking the phial and killing the cat. At any point the 'alive-or-dead' state of the cat can only be established by opening the box (the collapse of the wave function). According to the Copenhagen Interpretation at some point the cat is both alive and dead.

Einstein wrote to Schrödinger to approve this illustration of the absurdity of the Copenhagen quantum interpretation. Since then quantum theory has argued that macro events (like living and dead cats) were never part of the theory; it only applied to sub-atomic events. Very recently the entanglement between events has moved precisely to macro entanglement, with the suggestion that far from the observation (opening the box) collapsing the function, actually gives it 'coherence'. Observation 'creates' the reality (which suits Involution's hypothesis very pleasingly!) The entanglement of macro-events, and the 'reality' that is the result of collective perception is fundamental to the Theory of Involution. In the light of recent understandings of 'non-local hyper- communication' the Schrödinger cat can possibly be both alive and dead, since entanglement is increasingly seen as universal, and life and death could be reciprocally entangled at time zero, as would be thought and creation.

360   The oldest part of the brain is called the 'reptilian'; it controls the automatic somatic processes, such as breathing, heart beat, respiration, sweating. The serpent here is being caustic about being limited to an assigned 'scullery' at the base of the skull, whereas it speaks everywhere, in every cell and throughout creation.

361   **DNA reading the 'context' of its necessary interpretation through RNA's intermediacy and moderation. Laser light retaining 'information' from DNA and transmitting it.**
As mentioned, bacterial cells are not nucleated with a separated membrane, which accounts for their very rapid spreading of duplication and infection. Viruses work by utilising the host genetic material. Both contain RNA. As outlined above, it is thought that RNA was the less precise coding molecule that pre-existed DNA, but which was subsequently 'conscripted' to do the donkey work of protein synthesis. It is now thought by Garjajev et al that RNA plays a role in giving DNA its 'contextual' meaning, determining (from alternatives) the appropriate codons; so as 'messenger' it works in both

directions from and to DNA.

Prions are non-nucleated parasitic proteins that cause brain diseases like 'Mad Cow', Alzheimer's, and Kreutzfeld-Jacob Syndrome. They are highly strain specific, but without the DNA to 'recognise' the strain they infect, the puzzle is how do they detect their victim? It is now suggested that proteins can be 'read' by RNA through polarized photons and this constant updating of DNA's content has been moderated by RNA. If true then the simplest viruses and bacteria have helped to integrate the biocomputer of increasingly complex consciousness. Garjajev is now engaged in devising a polarisation-laser-radio-wave (PLRW) immunology. Through treating disease with modulated (by exposure to sound) laser waves the replication of infection can be interrupted by changing the 'context' in which replication is read, a sort of 'semantic noise' confusing the infecting virus. This 'corrected' DNA is then adopted by the host as a form of immunisation. The idea of talking seriously to DNA (and beating him when he sneezes) has begun.

362   When answering criticisms of his theory of evolution Darwin admitted, in later editions of *The Origin of Species*, that Nature was 'niggard in invention'. Here the serpent admits the slow pace of change in the early universe. Until complexity of action and the use of tools accelerated the speed of evolution, options on structure were limited by slow pace and simple forms. Yet now these simple forms are seen to retain capacities more complex ones have lost. (Planaria can regenerate from a part; slime moulds assemble when they need to reproduce.)

363   The Oracle of Delphi was deemed to interpret the message from the Gods. So too, here, does the serpent intercede.

364   The meninges, *the thick membranes of the brain* enclose the cortex. Here the serpent is suggesting these constitute the moat surrounding the citadel, and limiting its understanding of what lies beyond. The Styx was the river between life and death across which the boatman, Charon, ferried the soul.

365   Ouroboros, the symbol of the snake biting its tail, (Greek) Quetzalcoatl, the Feathered Serpent (Aztec) Sito the primordial

serpent (Egypt) Ronin, the two headed anaconda (Peru) Yin and Yang the twin elements intertwined (China, Japan) are all mythological representations of the creative principal and principle.

366 **Memory, holographically retained everywhere, may not be located in brain areas.**
The 100 year old history of attempting to locate, and apportion memory storage in the areas of the brain has failed spectacularly. Destruction of large brain areas by injury or surgery often has almost no impact on memory. Sheldrake quotes the case of a seemingly intelligent young man with an IQ of 126 with only 5% of normal brain structure. Karl Lashley spent thirty years trying to 'find' memories of various kinds and his student, Karl Pribram eventually supposed memory was stored as a hologram with each in all and all in each spread throughout the brain.

It would appear that DNA perhaps fits this bill better than any alternative, since it is the only chemical or cellular structure that is not replaced by decay and growth. Research on 'learning' in caterpillar larvae shows the persistence of the memory (of toxic exposure) in the butterfly that emerges after the complete liquefaction and reconstruction during the pupal phase. Sheldrake suggests this has to be due to Morphic resonance—the traces left by every other butterfly—somewhere. Involution suggests that the 'Morphic traces' would be memory, retained both within the specific DNA of the caterpillar butterfly (as in the larva which retains the memory of toxic exposure) as well as in the resonant DNA everywhere. But the long held assumption that mind is brain is almost impossible to shift. Even a thinker like Sheldrake does not posit another molecular structure which, like a magnet, could create a 'field', and, in harmony with others of its kind, a coherent field of vast and perhaps limitless dimensions, stretching both backwards in time, and forwards to mediate development. (See footnote 383 below)

367 The echo of Christ's admonition; 'Except ye be converted, and become as little children, ye shall not enter into the kingdom of heaven' (Matthew 18:3). The serpent extends this to the innocence from intellect's duality, and the deception of preconceived ideas or doctrines.

Involution

368   DNA— Deoxyribonucleic Acid— is defined by its lack of one oxygen, less than Ribonucleic Acid (RNA). A sort of irony since oxygen was the precondition for all life.

369   *'One transcript of our ladder'*. Since DNA can replicate itself from one strand, it puts a new complexion on the virgin birth. A male child would need a Y chromosome, or the absence of a second X, but not necessarily a 'father'.

370   Pheromones (from *Phero* (Gk) - to bear or carry + *Hormone* - impetus) are chemicals capable of acting outside the body of the secreting animal to trigger responses in others. Certain Lepidoptera (moths and butterflies) can, through sex pheromones, detect a potential mate at a distance of 10k (6.25 miles).

371   Petrarch (1304-1374) an Italian poet and scholar, known as the 'Father of Humanism', as well as the originator (with Dante and Boccaccio) of modern Italian, was originally forced to take up law by his father. He rejected 'making a merchandise of my mind' and devoted himself to literature and travel. The success of his work made him the first Poet Laureate since antiquity, after Homer and Cicero. He worked in Avignon for Pope Clement V and as a churchman was not permitted to marry.

In 1327 on Good Friday in church he caught sight of Laura, and she fired in him an unrequited passion that lasted his life (as Beatrice did for Dante). She inspired his writing of '*Il Canzoniere*', a book of 366 poems/songs. He wrote 'I struggled constantly with an overwhelming but pure love affair- my only one'. His romantic longings emerge from his evocation rather than description of her, never rising above her 'bel pié', her lovely foot.

So we return to the subjects of the first Canto, the perennial poets, and the romantic role of the female goddess to inspiration. Is it too soon to postulate that the disproportional contribution of homosexuals to the fine arts, particularly of great spiritual works by Michelangelo, Leonardo, Tchaikovsky, possibly also Shakespeare and many poets, might be due to the internalized female recognising union with the Divine? Yet also having the liberty and male energy to manifest and create when few women could?

372 **The universality of the Fibonacci Series; pi and phi.**
Fibonacci's Series, was named after Leonardo of Pisa (1207), called Fibonacci, who wrote a book *'Liber Abaci'* which introduced the series to the West. They were originally described in India by Pingola (200BC) or Virahan (AD700). The numbers are 0,1,1,2,3,5,8,13,21,34.....etc. Each is arrived at by the sum of the two previous, which bears out Plato's observation that 'when two things are bonded, it is by a third'. Trinities seem inescapable. Its adoption throughout the harmony of nature's patterns and growth was something that took centuries to recognise (because biological science was relatively late to evolve), but the beauty of its mental existence has been recognised since early thought. The golden section phi is achieved by this asymmetric relationship of each to each other and to the whole.

When a line is divided in such a way that the smaller segment bears the same ratio to the larger, as the larger does to the whole you arrive at phi. If each of the Fibonacci numbers is divided by the previous 1/0; 1/1; 2/1; 5/3; 8/5; 13/8, putting them in decimal form and plotting them on a graph, the harmonics of phi show as the form of the golden spiral which is adopted throughout natures' growth and relationships; from shells, to florets, to leaves, as well as natural harmonics in music. (Not only is Plato recovered, but also Pythagoras's intuition that number was all.)

Further, if the twelfth term of the series (89) is divided into unity, the decimal that results reproduces the same series 1/89 = 0.01123581321345589144… From the very large to the very small the language of number and its harmony runs throughout, as do 12 apostles, 12 solar months, 12 signs of the zodiac, 12 tribes of Israel, 12 notes of the octave scale. The numbers also trace the ancestry of honeybees (which produces drones) so that one male bee has one parent, 2grand, 3great grand, 5 great great etc. Perhaps sterility in bees was not created merely for industry (the neo-Darwinian supposition) but because number asserted its influence for the sake of all.

373 The golden angle is achieved by dividing a circle into two similar proportions, according to phi. The angle subtended by the smaller section's arc is 137.51. This is the angle most commonly adopted by nature's spiral growth because it never aligns one leaf above another. Each is exposed to the sun by incremental rotation. Although the DNA spiral does not resemble the golden in shape, it ultimately codes

for it in nature.

374 The language of DNA as the fundamental language of mathematics, music and the natural world.

'I have triangles for Pythagoras...'. Every *second* Fibonacci number forms the hypotenuse of a right angled triangle with integer sides. The length of this hypotenuse is equal to the sum of the three sides of the previous triangle but also, working backwards: The shortest side of any of these triangles is equal to the difference between the previous bypassed number's shortest side and the shortest of the previous triangle. So the first triangle in the series has sides of length 5, 4, and 3. Skipping 8, the next triangle has sides of length 13, 12 (5+4+3), and 5 (8-3). Skipping 21, the next triangle has sides of length 34, 30 (13+12+5) and 16 (21-5). This series continues indefinitely. The Pythagorean relationship refers both forwards and backwards simultaneously.

The beauty of mathematics has inspired creative mathematicians long before any application of their theorems or equations were needed. The serpent is suggesting that its language spoke directly to inspired minds, which is how mathematics 'found' itself in what had preceded the arrival of man—nature. DNA embodies all relationships underpinning the perfect (asymmetric) phi harmony of creation. Analysis of musical compositions seems to show an instinctive application of the golden mean in the proportions of symphonies, (Beethoven, Bartok) and even the use of harmonic relationships in phrasing, climax and resolutions. The sympathetic vibrations of notes an octave apart are recognised by mammals and some birds and seem deep 'embedded' in the thalamus of musicians, whose neurons in that area seem pre-tuned to acoustic harmony. The 'ur-alphabet' is now suggested as the four bases of which DNA is built. Further integration between Fibonacci's series and the actual arrangement of the microtubular lattice structure converting consciousness to structure is recorded in the two footnotes that follow (375 and 376).

375 The creation of structure from energy: computation and linguistic hypotheses.

Recent developments have started to address the role of consciousness in evolution, and in the conversion of ambient energy to structure. These are very recent and their arrival illustrates the processes of scientific understanding; the filling of loci into which theory is tending to fall. The first (by a mathematician, Penrose) details a method of mathematical computation (the 'how'); the second (by a bio-physicist, Garjajev), a linguistic and vibrational communication (the 'what', and the 'when'). No doubt both will be united ultimately.

A recent paper by Penrose and Hameroff puts forward a detailed proposal that explains brain function and Whitehead's 'drops of awareness' in terms of quantum processes through 'biologically orchestrated quantum computations' (Orch OR moments) in collections of microtubules within the brain neurons. The OrchOR hypothesis extends the standard quantum OR (Objective Reduction, the 'collapse of the wave function') understanding of quantum mechanics by linking a brain bio-molecular process to the fine structure of the Universe. The abstract of this paper ends with the statement *'We conclude that consciousness plays an intrinsic role in the Universe'.*

Not only does it bridge the gap between the mind and matter, (and unite the paradox that makes classical and quantum physics irreconcilable) but it does so by linking the mechanisms and structure of 'awareness' with actual biological cell structures. These are present in all eukaryotic cells, from the simplest unicellular organism to the integration of the complex human brain's neurons, where they work in coherent matrices of great numbers. The simplest cell (slime mould) shows an ability to find food, orientate or react to an imminent threat without any nervous system. This hypothesis offers an explanation which, if true, would support involution's supposed development of increasingly complex consciousness within organisms, using earlier simpler methods re-integrated and later collectively orchestrated.

Until recently, brain EEG's or gamma synchrony was thought to derive from the synaptic neuronal firings leading to the belief that thought or consciousness was the result of brain function (the emission image). This is now known not to be the case. Measurable brain activity often follows *after* a conscious response to a stimulus, which lethally undermines the idea of brain-as-processor and consciousness as its epiphenomenon.

Earlier, in the footnote on chromosome separation prior to cell division, mention was made of protein microtubules to which the centromere of the chromosome attached, in order to be pulled apart from its twin during preparation for cell division. Similar microtubules are now proposed to form lattices in all cells and to orchestrate complex behaviour and orientation, acting as a bio-molecular computer. Microtubules are self assembling polymers of protein (tubulin) which form proto-filaments aligning side-by-side to form such hollow tubules. These show two types of lattice; the A lattice has winding patterns which intersect the proto-filaments at intervals which match the Fibonacci series in a helical symmetry. This adherence to Fibonacci in the structural organization of a cellular lattice suggests a *'sympathy with the large scale processes found throughout nature'.* (Penrose) Garjajev offers further 'large scale' explanations below. Hameroff has shown that such microtubules establish cell shape, direct growth and organise functions of single cells and brain neurons.

Like DNA, microtubules fuse side-by-side in doublets or triplets and the alignment of nine (3x3!) form the cilia and flagellae that afford movement and spatio-temporal orientation in previously immobile cells. (So the tubules that permit a spermatozoon to fertilise an egg also perhaps permit the conversion of mind to matter.) Hamaroff has long suggested that these microtubules acted as information- processing cellular 'automata.' (continued in Footnote 376)

376 **Entangled consciousness as a quantum phenomenon.**
The 'units' of the lattice can exist in specific states (0 and 1) and each interacts with its neighbours in discrete steps, permitting complex computation. Earlier, Fröhlich suggested that such tubulin microtubules would oscillate synchronously, coupling or condensing, to a common vibrational mode, thereby converting ambient energy to coherent excitation. Penrose extends this by suggesting that such microtubule automata in brain neurons could be the basis of information processing, by organising synaptic transmission, learning and memory and be the *'substrate for consciousness'.*

The proposal of 'orchestrated objective reduction' (OrchOR, requiring coherence between many contributors) bridges the gap between classical physics (the world as it appears…after the OR) and the probabilistic quantum world. In the latter the equivalence of many possible OR's are reduced to the single OR by the measurement, itself a quantum system subject to the same uncertainties. Instead, orchestrated coherence (OrchOR) is no longer random but acts according to some non-random computational process, consciousness, but *'only when the alternatives are part of some highly organized structure and thereby acting in an orchestrated (Orch) form.'* Penrose goes further to suggest that a major component of the contribution to reaching Orch form would occur *'in the environment'*.

'Since the environment will be quantum entangled with the system (in the organism) the 'state reduction' in the environment will effect a simultaneous reduction in the system. This could considerably shorten the time for R to take place.' However it would also introduce the apparent randomness, **masking** the distinction between the ordinary R of quantum mechanics and the OR of biological systems. Such a synchrony or resonance between 'external' and 'internal' systems would provide a mechanism for the incremental growth of consciousness (involution) up the phylogenetic ladder as suggested. Such quantum coherence has been shown to occur in warm biological systems, in photosynthesis, bird brain navigation and other essential biological chemical processes. The shielding of the process by the lattice structure (described above) and the formation of hydrophobic pockets prevents quantum decoherence.

Addressing evolution in terms of this hypothesis, Penrose suggests that classical computation through micro-tubular cellular automata would have preceded the more sophisticated Orch OR process which required protection from the environment and the complex neuronal organisation which is found in higher animals. Such microtubules might have appeared in the eukaryotic cells 1.3 billion years ago, initially providing locomotive structures (cilia and flagellae), and when sufficiently complex structures provided the necessary protection from decoherence, and OrchOR replaced the more random OR (possibly at the time of the Precambrian explosion), evolution accelerated.

There is one further aspect to the quantum nature of consciousness that could support its forwards causative nature; the perception of time reversal, and 'backwards causality'. Penrose accounts for this by suggesting that quantum entanglement does not, over short periods, follow the normal sequential rules of causality… *'providing this does not lead to a contradiction with external causality.'* From this he concludes that *'accordingly OrchOR could well enable consciousness to have a causal efficacy.'* (i.e. a backwards causality) **If proved correct, then Orch OR suggests that conscious experience plays a role in the operation of the laws of our Universe.**

377   **RNA as messenger both to and from DNA?**
'Kinship writes that calling card'. It is in the 'junk' sequences of DNA that close relationships are detected. Those sequences show degrees of kinship, and are routinely used by forensic detection, paternity claims etc. Although Penrose's hypothesis above gives a coherent explanation of the 'how' whereby consciousness might create, it makes no attempt to explain the 'what' of memory. The relationship between microtubule OrchOR's and long term DNA storage will be the next field to conquer and connect together if involution's hypothesis is to be validated. In 1962 it was already known (Hayden see Schmitt) that learning in a mature organism resulted in permanent changes to the cell's RNA. Could RNA perhaps transfer not only the DNA's instructions to the cell's manufacture of proteins at the ribosomal 'bench', but equally transfer memory from cellular RNA to more permanent storage in the cell's DNA? (Rather like a waitress returning to the kitchen, but never empty handed.) Might this account for the differences between short and long term memory recovery, and how ageing affects it? (See below in Note 383.)

378   In recent contradiction to this statement see Notes 374 and 375.

379   A reference to Henri Bergson who said that *'The intellect may be compared to a carver, but it has the peculiarity of imagining that the chicken always was the separate pieces into which the carving knife divides it.'*

380   *Putzlappen-* dishcloth or duster. A *Putzmädchen-* cleaning girl.
'Too many notes, Herr Mozart' was allegedly said by Prince Leopold after a Mozart performance.

381   The serpent is suggesting that the adolescent intellect (*primping in the mirror-*mistaking the nature of its reflection), by failing to see the immediacy of creation, creates the causal delay between thought and act…that which we recognise as 'time' *'its older sister's clothes'—* outworn understanding. Penrose (Note 376) acknowledges the causal nature of consciousness that can work backwards in time since quantum entanglement does not follow the usual sequential progression of classical physics and consciousness is proposed to be a quantum phenomenon. Instant creation (thought as act) is now

implied as true beyond the limited context of the quantum world, (although perhaps mediated by such connections in the brain) is here suggested for all inspired 'subito' connections in the eurekas of genius.

382   The dawn *'reveille'* is this dawning new call to deeper understanding. The 'last post' is the suggestion that science only attends to a single life, as the field wherein it works. Imagine medical research tempered to a belief in re-incarnation? Or neuroscience tempered by a causative biological memory of the past? Or the law on euthanasia governed by a belief in an alternative healthy body? Now we really have, in science, approached the completion of the old journey of recovery and stand on the brink; the new journey into the interconnectedness of consciousness. (To whet anticipation, see following note.)

383   **DNA as a bio-computer and non-local communications network.**

The supposition underpinning this Theory of Involution in 1970 was the existence of a method of memory storage and retrieval, and originally an intuitive guess that the sequences of 'junk' DNA, the 95+% not used for protein manufacture had the required complexity and longevity to provide it.

The work of the Russian bio-physicist Pjotr Garjajev (Peter Gariaev) now gives cogent support to the language in junk DNA as the archetypal cellular language, underpinning and permitting the acquisition of all the human tongues: mirroring words, syntax, grammar, logic and rules, as well as homonyms (whose meaning is context dependent as are homonyms such as wet and whet). Not only is DNA the earliest language; it responds directly to modulated human languages and sounds, which explains the power of hypnotic suggestions, and elevated states of consciousness brought about by chanting and mantras. It also now may account for clairvoyance, telepathy and other forms of hyper communication. The power of such language has long been known to the esoteric and hermetic traditions.

Instead of approaching DNA as a chemical structure to be cut and spliced (the Western way…kill it first) the Russians examined it as a language to be deciphered and understood. What they have discovered will revolutionise almost everything. The Watson-Crick-Chargaff encoding of the triplet codons (elaborated above for the making of proteins) requires an additional flexible tongue, a microscopic equivalent of the EPR (Einstein-Podolsky-Rosen quantum bridges) for the transmission of information. Here is where the material 5% may be shown to be shaped by the 95% non-material, the music of history. The Russians have demonstrated that modulated laser light (by exposure to DNA) can repair DNA damaged by x rays, such as in seeds irradiated by Chernobyl, as well as transferring codons from one form of organism to another. The responsiveness to sound waves and the information retained by a laser's encounter with DNA, even after the DNA was removed-(Garjajev and later Poponin) indicated that junk DNA appears to co-ordinate all information required for the integration of biological systems. These non-coding areas had more in common with the long distance correlations present in spoken language than the coding areas which only need to assemble proteins.

384   **DNA as the biological internet.**

In reporting this work (originally published in Russian), Fosar and Bludorf state *'In truth DNA is not just a blue print for constructing the body, it is a storage mechanism for optical information as well as an organ for communication'*. Following the discovery that biophotons are constantly emitted at a resonating frequency of 150 MHz (F.A.Popp) Garjajev describes how DNA not only receives and transmits electromagnetic radiation but it *'absorbs the information contained within the radiation and interprets it…a complex interactive optical biochip'*

'Human DNA is a biological internet. It acts as a data storage and communication system.' As it is the underlying language from which others were later derived, it is *'normal and natural for DNA to respond to language…no decoding is necessary.* (Perhaps Sheldrake's dogs do not need to meet the statisticians after all.)

This work greatly clarifies the mechanisms whereby 'Morphogenetic fields' may integrate and differentiate embryological development in which, from the same DNA in every cell, different cells and organ systems develop in sequence and differentially, as leaves, as stalks, as eyes or liver. It also allows for the constant updating or improvement from experience.

The hyper-communication between DNA was analysed and it was found it followed complicated non-linear wave formations called soliton waves, which are stable and able to store information over long periods of time. Chromosomes work as 'solitonic holographic computers' under the influence of endogenous DNA radiation. So the junk, often called 'silent' is far from silent, it chatters constantly. Since this is so critical a piece of the puzzle I quote Fosar and Bludorf's translation of Garjajev: He states:-

1) *The evolution of bio-systems has created genetic 'texts' similar to natural context-dependent texts in human language.*
2) *The chromosome apparatus acts simultaneously as source and receiver of these genetic texts- decoding and encoding them.*
3) *The chromosome continuum is analogous to a static-dynamical multiplex time-space holographic grating (the space time of an organism).*

DNA is a gene- sign laser and its electro-acoustic fields are <u>read</u> by the gene bio-computer as a human would read a language text. These texts form 'holographic <u>pre-images</u> of bio-structures and the organism as a whole (dynamical wave-copiers…this is the field for <u>calibrating</u> the bio system).

385 **Non-local or hypercommunication, the interrogation of matter by DNA. The structure of the 'book' of DNA, contextually dependent.**

A critical component of this explanation is due to the *non-local* nature of DNA information, (everywhere at once, changing simultaneously). This is shown throughout all the levels of development, becoming increasingly complex. Specialisation is accompanied by integration. The process moves from organism-as-a-whole (as in Planaria which can regenerate from part) through the cellular self protection (immunity, integrity), to the cell nucleus (synthesizing, repairing), the molecular (sequencing and preservation) and finally escaping the organism completely in the chromosomal holography which operates throughout the biosphere. This total integration of messengers and messages has been implicit in the perfection of intricate creation and its progress, to reveal through memory and collective mind its own 'holism'. Man's mind is the product of this incremental collective integrated sophistication.

Chromosomes radiate and also 'transmit and read' by transforming radiated photons into a 'broadband of genetically signalled radio waves'. Somehow a system of laser mirrors, polarizing emissions, can be 'read out' by iso-morphous (same shape) radio waves. The image is dynamic, depending on the matter *interrogated* (by the DNA). This is an entirely new form of computer, a bio-computer and a new type of EPR spectroscopy which conveys the features of different objects, stored by laser mirrors and recurring in time. Garjajev calls this *'a storage/recording environment…capable of directly recording the space-time behaviour of objects…the basis of a new type of video, audio recording or cinema'* (Our artists and dreamers have been posited to have viewed this cinema often, and time travel has preoccupied science fiction for centuries, now perhaps consciousness will read 'historic DNA'.)

Garjajev has validated this capacity of DNA to 'read' polarised photons. By exposing potatoes to modulated laser radiation (modulated by its exposure to DNA) this irradiation produced greatly accelerated growth, and tubers developed on stalks, not only the roots. Controls with lasers unexposed to DNA had no effect. The normal development of tubers on roots only, was confused by the stalks subjected to 'root-like' DNA instruction. This 'Phantom DNA', seemingly is retained by the laser exposed to it, and is stored non-locally. Interestingly, as support for the hypothesis of involution throughout history, it has long been known that, although macromolecules (DNA, RNA Proteins) have an outspoken capacity to an *optical rotatory dispersion of visible light,* so do smaller micro-molecules (saccharides, nucleotides and amino acids) which has never made biological sense. Its value, if any, was unknown.

Perhaps, as an early stage 'precursor', this 'quantized optical activity' (whereby the organism obtains 'unlimited information on its own metabolism') will provide the equivalent of the Higgs boson as the ur-mechanism for the acceleration, integration and complexification of consciousness and evolution. The slime-mould was mentioned right at the start of this history of consciousness, and the ability of simple organisms to repair themselves from just a part (Planaria and earthworms) is now part of the evolving continuum. The loss of that capacity may now be reparable just by talking nicely to DNA. (Perhaps in Russian.)

So to conclude the technically dense but indicative evidence: - DNA is shown to be a self-calibrating antenna working by adaptive resonance, capable of both receiving and transmitting quantum holographic information stored as diffraction patterns (holograms).

Morphic resonance during embryological development is now understood as the reading and co-ordination of this information to produce the perfection of each organism and each part, at the right times, in the right order.

Earlier footnotes described the anti-parallelism of the two mirrors strands of DNA. This is now shown as critical to the quantum hologram information storage which requires the two *anti-parallel* helices *between which* are located the planes of the hologram where the three spatial dimensions and image data are stored. We have seen the base pairings between A-T, G-C: These are kept separated by the anti-parallel helices (which, although mirror images, are phase conjugate) thereby constituting 'pages' of associative holographic memory which ensure they are not 'cross read'. So time sequencing as well as content is critical. The 'pages' of the DNA book gives the context, enabling a varied language from a limited vocabulary. So DNA not only talks in words; it writes in context. (These footnotes are derived from very recent work, which no doubt, will be further clarified and validated.)

386   Peter Abelard (1079-1142) was a French scholar, philosopher and theologian, who became a renowned teacher and then professor at Notre Dame, Paris. The Canon of Notre Dame's niece was Hélòise, both beautiful and classically educated. Unlike Petrarch, Abelard joyfully seduced her; she became pregnant and was sent away to have her child who she called 'Astrolabe' (after the instrument). Eventually they were permitted to marry but Hélòise rejected marriage and took to a nunnery. Abelard was castrated, chasteness forced upon him. The two were widely known in illustrated copies of the 'Roman de la Rose'. Hélòise eventually became an abbess at a convent under Abelard's jurisdiction where they continued to meet. After he died his remains were taken to her and later she was buried with him. The famous romantic lovers were transferred to a splendid vault in Père-Lachaise Cemetery, Paris.

The remainder of this Canto celebrates the passionate love of all romance, and the mystic marriage of love in the Divine. Although love is perhaps the integrated 'field' of all Creation, it is only the individual that can experience its transpersonal nature by surrender of the self. Romantic obsession and absorption are the ordinary human approximation of the mystic's experience. The glorious paradox is that those who have experienced such self-forgetful love, become more expressive as individuals, (*'plucked from brittle anonymity'*) drawing others towards the light by languages made unique but echoing the perennial. The methods of creation continue in human creativity.

Bibliography

This work is a very personal reflection drawn from a lifetime's encounters with books, conversations, and changing interests, not academic or narrowed to any field of expertise. As a result, this bibliography is far from comprehensive, nor could it be. I have attempted to provide the references for specific points mentioned, only to strengthen the journey of involution and to provide its vocabulary and connections. Many date back to a preliminary education in the 1960s; several are standard texts, and a few are recent corroborations in an eclectic mix of subject areas. Another collection would do just as well, but these were what fell my way.

Ambrose, E.J., *The Nature and Origin of the Biological World* (New York: Halstead Press, 1982).
Aurobindo, Sri, *The Adventure of Consciousness* trans. by Tehmi (Pondicherry: Aurobindo Ashram Press, 1968).
Bartholomew, A., *Hidden Nature: The Startling Insights of Viktor Schauberger* (Edinburgh: Floris Books, 2003).
Beveridge, W.I.B., *The Art of Scientific Investigation* (London: Heinemann, 1950).
Böhm, D., *Wholeness and the Implicate Order.* (London: Routledge and Kegan Paul.1980).
_____& Hiley, B., *The Undivided Universe* (London: Routledge, 1993).
Bragg, Sir Lawrence, *The History of Science* (London: Cohen & West, 1951).
Brodie, R., *Virus of the Mind: The New Science of the Meme* (Seattle: Integral Press, 1996).
Bronowski, J., *Science and Human Values* (London: Hutchinson, 1961).
Burnet, J., *Early Greek Philosophy* (London: Blackwell, 1908).
Butterfield, H., *The Origins of Modern Science* (London: G. Bell, 1949).

Calladine, C.R., & Drew, H., *Understanding DNA.* (London: Academic Press, 1992).
Casti, J.L., *Paradigms Lost: Images of Man in the Mirror of Science* (London: Scribners, 1996).
Chardin, Pere Teilhard de, *The Phenomenon of Man* (London: William Collins, 1955).
Clark, K., *Civilisation* (London: BBC, 1969).
Collins, F., *The Language of God* (London: Simon and Schuster, UK. 2007).
Conway Morris, S., *The Crucible of Creation* (Oxford: OUP, 1998).
Crick, F., *Life Itself: Its Origin and Nature* (London: MacDonald & Co, 1981).
Dante. *The Divine Comedy*, trans. C.H. Sisson (Oxford: OUP, 1980).
Darwin, C., *On the Origin of Species by Means of Natural Selection* 6th Edition (New York: Appleton & Company, 1898).
Dawkins, R., *The Selfish Gene* (Cambridge: Cambridge University Press, 1976).
The Extended Phenotype (Oxford: W.M. Freeman, 1982).
The Blind Watchmaker (Harlow: Longman, 1986).
The Theory of Evolution (Cambridge: Cambridge University Press, 1993).
River Out of Eden (London: Weidenfeld & Nicolson, 1995).
Unweaving the Rainbow (London: Penguin, 1998).
The God Delusion (London: Bantam Press, 2006).
Dilger, W.C., 'The Comparative Ethology of the African Parrot, Genus Agapornis', in *Zeitschrift für Tierpsychologie* Vol 17, 649–685, 1960.
deWaal, F.B.M., *Good Natured: The Origin of Right and Wrong in Humans and Other Animals* (Cambridge, MA: Harvard University Press, 1996).
Bonobo: The Forgotten Ape. (Berkeley: University of California Press, 1997).
The Age of Empathy (New York: Harmony Books, 2009).
Dingle, H., *The Scientific Adventure: Essays in the History and Philosophy of Science* (London: Pitman and Sons, 1952).
Dobzhansky, T., *Genetics and the Origin of Species* (Columbia: University Press, 1982).
Einstein, A., Podolsky B. & Rosen, N. 'Can quantum-mechanical description of physical reality be considered complete?' in *Phys.*

Rev. Vol. 47 (1935),777–780

Eldridge, N. & Gould, S.J. 'Punctuated Equilibria: An Alternative to Phylogenetic Gradualism' in *Models in Paleobiology*, (ed.) Thomas J.M. Schopf. (San Francisco: Freeman Cooper, 1972).

Eliot. T.S., *Selected Poems* (London: Faber & Faber, 1954).

Empson. W. *Seven Types of Ambiguity* (London: Chatto & Windus, 1930).

Emery, N.S. & Clayton. N.S., 'The Mentality of Crows: Convergent Evolution of Intelligence in Corvids and Apes' in *Science* 306 (2004), 1903–1907.

Feynman, R., *The Character of Physical Law* (London: BBC, 1965).

Fraenkel, G.S. & Gunn, D.L., *The Orientation of Animals* (Oxford: Clarendon Press, 1940).

Frisch, K. von, *The Dance, Language and Orientation of Bees* (New York: Ithaca, 1950).

Fosar, G. & Bludorf, F., *Vernetzte Intelligenz.* (Dusseldorf: Omega Verlag 2011)

Gardner. E., *Fundamentals of Neurology* (Philadelphia: W.B. Saunders & Co, 1959)

Gardner. H., *Art Through the Ages* (London: G. Bell & Sons, 1927)

Gariaev P., & Tertishniy, G., 'The quantum non-locality of genomes as a main factor of the morphogenesis of biosystems', 3rd Scientific and Medical Network continental members meeting, Potsdam, Germany, May 6–9, pp. 37–39.(1999)

Gell Mann M., & Neuman, Y., *The Eightfold Way: Frontiers in Physics (*New York: Benjamin, 1964)

Gesell, A., *The First Five Years of Life* (London: Methuen, 1954).

Gombrich, E.H., *The Story of Art* (London: Phaidon Press.1962)
 Art and Illusion (London: Phaidon Press, 1964)

Gould, S.J., *Ontogeny and Phylogeny* (Cambridge, MA: Harvard University Press, 1977).
 _____ & Eldredge, N., 'Punctuated Equilibria: the tempo and mode of evolution reconsidered', *Paleobiology* 3, 1977.

Green, C., *Lucid Dreams* (Oxford: Institute of Psychophysical Research, 1968).

Greyson, B., 'Incidence and correlates of near-death experiences in a cardiac care unit', *Gen. Hosp. Psychiatry* 25, (2003), No 4.

Grof, S., *The Adventure of Self Discovery* (Albany: State of New York University Press, 1988), *The Holotropic Mind* (San Francisco: Harper, 1993).

Involution

    The Cosmic game: Explorations at the Frontier of Consciousness (Albany: Albany State University Press, 1999).

Hadamard, J., *The Psychology of Invention in the Mathematical Field* (Princeton: Princeton University Press, 1949).

Hall, D.B., et al. 'Oxidative DNA damage through long range electron transfer', *Nature* 382, (1996), 731–735.

Hardy, A., *The Living Stream* (London: Collins, 1965)

Hawking, S., *A Brief History of Time* (London: Bantam, 2003).

Hebb, D.O., *The Organisation of Behaviour* (New York: Wiley, 1949).

    _____*A Textbook of Psychology* (London: Saunders, 1951).

Heisenberg, W., *Physics and Philosophy* (New York: Harper Row, 1985).

Hilgard, W., *Theories of Learning* (London: Methuen, 1958).

Hilgard, E.R., & Marquis, D.G., *Conditioning and Learning* (London: Methuen, 1961).

Hinde, R.A., *Animal Behaviour* (New York: McGraw Hill, 1966).

Hofstadter, D.R., *Gödel Escher Bach: An Eternal Golden Braid* (London: Penguin, 1980).

Hoyle, F., *The Intelligent Universe* (London: Michael Joseph, 1983).

Hudson.JJ et al. Improved Measurement of the Shape of the Electron. *Nature* 473 (7348) (2011) 493–496

Hunt, G.R., 'Manufacture and use of hook-tools by New Caledonian crows', *Nature 379* (6562): 249–250, 1966.

Huxley, A., *The Perennial Philosophy* (New York: Harper, 2003).

Huxley, J., *Evolution: The Modern Synthesis* (London: Allen & Unwin, 1963).

    _____*Essays of a Humanist* (London: Chatto & Windus, 1964).

James, W., *Varieties of Religious Experience* (Abingdon: Routledge Classics, 2008).

Jung, C.G., *Psychology of the Unconscious* (New York: Moffat Yard, 1916).
    Modern Man in Search of his Soul (London: Kegan Paul, 1933).
    The Integration of Personality (London: Kegan Paul, 1940)
    Man and His Symbols (London: Aldus Books, 1964).
    Memories, Dreams, Reflections (New York: Random House, 1965).

Kingsley, P., *In the Dark Places of Wisdom* (Shaftesbury, UK: Element Books,1999).

    _____*Reality* (The Golden Sufi Center, 2003).

Koestler, A., *Insight and Outlook* (New York: Macmillan, 1949).

    _____*The Sleepwalkers* (London: Hutchinson, 1959).

    The Act of Creation (London: Hutchinson, 1964).

Köhler, W., *The Mentality of Apes* (London: Pelican Books, 1957).
Lashley, K.S., *Brain Mechanisms and Intelligence* (Chicago: Chicago University Press, 1929).
Laski. M. *Ecstasy: a Study of Some Secular and Religious Experiences* (London Thames and Hudson (1980)
Laszlo, I., *Science and the Akashic Field* (Rochester: Inner Traditions, 2004).
Science and the Reenchantment of the Cosmos (Rochester:: Inner Traditions, 2006)
The Akashic Experience (Rochester: Inner Traditions, 2009).
Lennox, J., *God's Undertaker: Has Science buried God?* (Oxford: Lion Hudson, 2007).
Lewis, C.S., *Rehabilitation and other Essays* (Oxford: Oxford University Press, 1939).
Lieber, M., Environmental and Genetic factors Affecting Chromosomal Instability at Mitosis in Aspergillus nidulans and the Importance of Chromosomal Instability in the Evolution of Developmental Systems. Evolutionary Theory vol. 1 (1975) pp 97-104
_____ The Effects of Temperature on Genetic Instability in Aspergillus nidulans. Mutation Research vol 34 (1976) pp 93-122
Livingstone, R.W., *The Legacy of Greece* (Oxford: Oxford University Press, 1951).
Lorenz, K. *The Waning of Humaneness* (Boston: Little Brown, 1987).
On Aggression (London: Methuen, 1966).
The Evolution of Behaviour (Scientific American, 1958).
'Phylogenetischen Anpassung und Adative Modifikation des Verhaltens', *Seits fur Tierpsychol 18,* 139–187, 1961.
Maniotis, A., Bojanowski,K., Ingber, D., Mechanical continuity and reversible chromosome disassembly within intact genomes removed from living cells. J. Cell. Biochem. Vol. 65 (1997) 114-130.
Margulis, L., & Sagan, D., *Microcosmos: Four billion years of Microbial Evolution.* (London: Allen & Unwin, 1987).
Masaru, E., 'The Message from Water iii', *Love Thyself* (London: Hay House, 2006)
Marcer, P., & Schempp, W., 'A Mathematically Specified Template for DNA and the Genetic Code, In terms of the Physically Realizable Processes of Quantum Holography, Proc. Greenwich Symposium

of Living Computers, (ed.)
Fedorec, A., & Marcer, P., 45–62, 1966.'Model of the Neuron working by Quantum Holography', *Informatica* 21. 5. 19–534, 1997.

Mayr, E.. *Principles of Systematic Zoology*. (New York: McGraw Hill, 1969).

McDougall, W., *The Group Mind* (New York: Putnam, 1920).

McGilchrist, I., *The Master and his Emissary: The Divided Brain and the Making of the Western World.* (Yale University Press, 2009).

McKenzie, A.E.E., *The Major Achievements of Science* (Cambridge: Cambridge University Press, 1960).

Medawar, P.B., *The Limits of Science* (Oxford: Oxford University Press, 1984).

Merton, R.H., *The Sociology of Science: Theoretical and Empirical Investigations* (Chicago: University of Chicago Press, 1973).

Narby, J. *The Cosmic Serpent* (London: Phoenix, 1999).

Ogburn, W.F. & Thomas, D. 'Are Inventions Inevitable? A Note on Social Evolution', Political Science Quarterly, Vol. 37, No. 1, 83–98, 1922.

Penrose, R., *The Road to Reality* (London: Random House, 2004).
   Shadows of the Mind: A Search for the Missing Science of Consciousness (Oxford: Oxford University Press, 2000).
   _____ & Hameroff, S., 'Consciousness in the Universe. Neuroscience, Quantum Space, Time Geometry and OrchOR Theory', Journal of Cosmology Vol 14, 2011.

Pinker, S., *How the Mind Works* (London: Allen Lane, 1997).

Playfair, G.L. *Twin Telepathy* (London: Vega: 2002)

Popp, Fritz–Albert et al (eds.) *Recent Advances in Biophoton Research and its Applications* (Singapore: World Scientific, 1992).

Popper, K.R., *The Logic of Scientific Discovery* (London: Hutchinson, 1959).

Rattemayer, M., et al, 'Evidence of photon emission from DNA in living systems', Naturwissenschaften 68, 1981.

Robinson, A., *Sudden Genius?* (Oxford: Oxford University Press, 2010).

Russell, B., *Why I am not a Christian* (London: Routledge, 1957).
   Religion and Science (Oxford: Oxford University Press, 1997).
   A History of Western Philosophy (New York: Simon & Schuster, 1945).
   An Outline of Philosophy (London: G Allen & Unwin 1927)

Sagan, C. *Intelligent Life in the Universe* (New York: Emerson Adams,

1966).

Scarre, C., (ed.) *The Human Past.* (London: Thames & Hudson, 2005).

Schmitt, F. O., (ed.) 'Macromolecular Specificity and Biological Memory', *Science.* 849 (Cambridge, MA: M.I.T. Press, 1962).

Schrödinger, I. *What is life?* (London: Macmillan, 1946).

My View of the World (Cambridge: Cambridge University Press, 1964).

Schütz, E., 'Die Spät-Auflassung ostpreussischer Jungstörche in West Deutschland durch die Vogelwarte Rossitten 1933', Vogelwarte 15, 63–78, 1949.

Schumacher.E.F. *Small is Beautiful- A Study of Economics as if People Mattered.* (London: Blond and Briggs.1973).

Schwaller de Lubicz, R.A., *The Egyptian Miracle: An Introduction to the Wisdom of the Temple* (Rochester: Inner Traditions, 1985).

Sheldrake, R., *The Presence of the Past* (London: Fontana Collins, 1988).

The Science Delusion (London: Simon & Schuster, 2012).

Sherwood T.F., *Science Past and Present* (London: Heinemann, 1949).

Singer, C. *A Short History of Science* (Oxford: Oxford University Press, 1941).

A Short History of Scientific Ideas to 1900 (Oxford: Oxford University Press,1959).

Stevenson, I., *Children who Remember Previous Lives* (Charlottesville: Charlottesville University Press, 1987).

*Twenty Cases Suggestive of Reincarnation (*New York: American Society for Psychical Research, 1966).

Thom R. *Mathematical Models of Morphogenesis.* (Chichester, Ellis Horwood, 1983).

Thorpe, W.H., *Learning and Instincts in Animals* (London: Methuen, 1956).

Tinbergen, N., *The Study of Instinct* (Oxford: Clarendon Press, 1951).

Turnbull. H.W, et al, eds. *The Correspondence of Isaac Newton* 7 Vols (Cambridge: Cambridge University Press 1959-1981).

Underhill, E., *Mysticism: The Study of the Nature and Development of Man's Spiritual Consciousness* (New York: Dutton, 1961).

Van Schaik, C.P., *Among Orangutans: Red Apes and the Rise of Human Culture* (Cambridge, MA: Belknap, 2004).

Virgil, *Georgics* Trans. C.Day Lewis *(*Readers Union, Letchworth, 1943).

Whitehead, A.N., *An Enquiry Concerning the Principles of Natural Knowledge* (Cambridge: Cambridge University Press, 1919).

Process and Reality (Cambridge: Cambridge University Press, 1929).

Wilbur, K., *Sex, Ecology and Spirituality: The Spirit of Evolution* (Boston: Shambhala Inc, 2000).

Wittgenstein, L., *Philosophical Remarks*, (ed.) R. Rhees trans. Hargreaves & White (Oxford: Blackwells, 1971).

Wolpert, L., *Six Impossible Things Before Breakfast: The Evolutionary Origins of Belief* (London: Faber & Faber, 2006).

Woolger, R., *Other Lives Other Selves* (New York: Doubleday, 1987).

Previous Publications

A Shadow in Yucatán

Reviews.
Alison Jakes (Poetry Circle)

'I was utterly awestruck by the writing skill and breadth of imaginative evocation…..poetic, elegiac…almost unbearably intense…sensuous imagery from both nature and modern urban living…musical, both rhythmic and assonant…sustained dramatic tension within a simple everyday story….the superficiality of the beauty salon is a very potent metaphor….'

Prunella Scales (Actress)

I think the writing is brilliant, she's obviously a very considerable poet…'

Katherine Knight (Real Writers)

As with a highly literary novel, this ambitious story makes demands upon its readers. As with most modern poetry it deserves to be read and re-read…..

The basic narrative is simple but poignant, taking a young unmarried girl through her pregnancy. She decides on, but then refuses abortion, is let down by her own mother but befriended by her landlady. She gives her child up for adoption and goes through a breakdown in consequence. The resolution is quietly sad, when she comes to terms with her situation, having seen her child loved by the adoptive mother.

The story is a vehicle for some impressive poetry. It is highly emotional and transforms the ordinary protagonist into an archetypal figure of suffering motherhood.

Involution

'*Speech must now grow from silence and the stones that cockle the black backs
Of women in pre-history, left alone with the consequence of men*'

It is not merely feminist polemic however, there is an impressive section 'The Storm' which celebrates the male principle in mythological terms, the storm God as fertiliser of the Earth, or the sky god as progenitor of the Goddess's son. There is religious dimension too. Throughout there are subtle references to the Christian Nativity, and on another level it tells of Christ's birth and Mary's suffering in modern terms. It contrasts the cruelty of the girl's Catholic mother, with the compassion of her Jewish landlady.

Considered as pure poetry, the lyrical passages show an accomplished writer at work. 'Light plaited the stalks into shadow, the dark rolled the sun in its sleeve, the wind took her pulse with a turning leaf, are merely examples. Sound and sense are well married and many devices such as alliteration are used skilfully. Occasional rhymes are unobtrusive. The varied line makes the verse as easy as possible to read given the length of the work…

There is implicit criticism of the hypocrisy of society as a whole….The poem has a social purpose.

Sarah Gorham (Editor in Chief, Sarabande Books Louisville KY)

I was so pleased to receive your fine collection, *A Shadow in Yucatan*. It's not often we get to see what British poets are up to. The poems are extremely skilful and quite moving. I like particularly your long lines and good ear; the latter is all too rare these days!'

'A welling unimpeded view of everything' –
　　　　from *The Recusant, On line Ezine. Editor Alan Morrison.*

Philippa Rees is as an immediately distinctive and striking poet who writes with unfashionable but often brilliant painterly verbal play and colour, oozing with a sensuous love of language – as opposed to many of her contemporaries who seem have a deep distrust of our vernacular, rather negating their own purpose in the process. Rees's almost tangible style frequently dazzles with imagistic chiaroscuro; stark contrasts of light

Involution

and shade, subtext and texture:

*Two saffron monks, pale-pated, cross the grass.
Their discourse falls in folds, their hands elaborate…*

*…The shafted pencil-light writes clearly on their
crowns; the ankles trace the shadows, but the bare
feet laugh…*

Breathtaking. This ripeness of verbiage and intrinsic musicality inevitably bring comparisons with Dylan Thomas (particularly the densely descriptive, rumble-tumble passages of *Under Milk Wood*):

But this is not to detract from Rees's individuality, which, throughout this book of poetic prose – a poetic narrative interspersed with verbalesque – is palpable and often beguiling. She is also prone to the lingering aphorism

'*The cradle of compassion lies in an open palm*'. And '*Nights are cloth soup silence*'.
and; '*There is no hook to recover resolve from the bottomless pit of despair*'
.and the unforgettable image – sometimes oblique, but still workably so:
'*Lethargy, that toothless crone, skims perpetual indifference from the cream of richer care*'.

Her poetic prose is punctuated by an irrepressible anthropomorphism; an instinctive gift at metaphorical personification. No object is permitted inanimate impunity in Rees's naturalist, rocking wordscapes:

The challenge of this 109 page piece is in absorbing and appreciating both its poetry and storyline at the same time, though ultimately they are as counterbalanced as deftly as one might hope from such an ambitious venture. For my part, I read *A Shadow in Yucatán* mainly for its poetry, its play with language, image and sound, rather than strictly trying to follow the actual narrative. Approaching this book with a cerebrally-dislocated Negative Capability, I experienced it in terms of descriptive impression, verbal effect. In this respect, *A Shadow in Yucatán* is disarmingly beautiful.

The Author: Philippa Rees

Shooting for the Pot.

The uninvited experiences that generated Involution have been covered in the Afterword but the life of extremes that provided the language seemed to feel its own way, never any compromises, it was always all-or-nothing. Solitude from early childhood in South Africa, (imprisoned in boarding schools from the age of six), but also the ecstasy of Noel, saddled at the gate on Christmas morning, and the limitless mountains of Lesotho with a saddlebag and no prohibitions; or on long safaris with a beloved and multilingual grandfather devoted to African education (and hers). First mathematics lay in calculating the frequency of fever trees to the mile, or the predicted girth of a Baobab, and language in finding rhymes for tsongalorla (millipede- Zulu) on the slow safaris visiting remote schools and camping under the stars. Snakes, interestingly, were the single terror. Elizabeth Barrett Browning, a thrice and well removed aunt, may have cast the spell (of independence) or perhaps her curse (of narrative poetry and its loquacious economy).

Ideas, to make sense of it all, were found in books, and the inspiration of two passionate teachers, of literature and theology. Choice was somehow never offered and when it was, everything was always better than something. At university five faculties were sampled before somehow arriving at science. Later life involved the Max Planck Institute in Bavaria, then Florida, four daughters, lecturing to mature university students, learning the cello (much too late), building an arts centre and concert hall, and living in Somerset, which continues. She writes constantly to find out what she thinks, but readers would clarify whether there is any value in it for anyone else. That would be new and marvellous. Other subordinated and shelved works wait in the wings.

The Author would welcome comments, questions and dialogue from readers, and can be reached by email at:- philipparees@btinternet.com

'Involution- An Odyssey Reconciling Science to God' and a previously published poetic novella, *A Shadow in Yucatán* can be obtained from Amazon (http://www.amazon.co.uk/Shadow-Yucatan-Philippa-Rees/dp/ or from her website (signed) at http://philipparees.wordpress.com/ or
http://www. collaborartbooks.com or
http://www.involution-odyssey.co.uk
http://www.involution-odyssey.com

CPSIA information can be obtained at www.ICGtesting.com
Printed in the USA
BVOW07s0041301014

372887BV00005B/404/P

'...it actually reflects the very act of genius.....

... reintroduces the aesthetic, beautiful, meaningful process that is poetry into science.

...clearly written and easy to read.'

Philip Franses: Editor, Holistic Science Journal

'A brilliant and profoundly erudite epic...

The author's grasp of the principal elements of Western culture is masterly and her poetic narrative woven together with extraordinary subtlety.

...a heroic intellectual tour de force...deserves the widest readership.'

David Lorimer: Director, Scientific and Medical Network

*
* *

Involution: Mavericks and Inspiration.

Two companions, Reason and Soul, embark upon a light-hearted poetic journey through creation's evolution, and its legacy: memory. Memory's recovery is traced through the intuitions of scientific genius: from pre-Socratic Greece, through the epochs of Western thought to the dissolution of Modernism, where mind and matter re-approach their recovered and essential unity.

Adding involution to evolution is adding yeast to flour; it bakes a lighter loaf. Saints and scientists break the same bread.

What is revealed is a science beyond, one that honours its prophets and restores man to his origins, his deeper self.